"There are millions of unfortunate people suffering from mold toxicity, and most of them do not even realize that mold toxicity exists as a common cause of illness. Laura Linn Knight has written an excellent discussion of this medical condition, using the experience of her own family and others, to bring this important realization to life. She helps the reader to understand what mold toxicity is, and how to diagnose it and treat it, which will be an invaluable guide for many who are suffering with this condition. She incorporates her background of meditation into practical methods to cope and deal with the illness and the anxiety and depression it can trigger. This is an excellent introduction to this often underdiagnosed medical problem."

—Neil Nathan, MD, author of *Toxic: Heal Your Body from Mold Toxicity, Lyme Disease, Multiple Chemical Sensitivities and Chronic Environmental Illness* and *The Sensitive Patient's Healing Guide*

THE
TOXIC
MOLD
SOLUTION

A COMPREHENSIVE GUIDE TO
HEALING YOUR HOME AND BODY FROM MOLD:
FROM PHYSICAL SYMPTOMS TO
TESTS AND EVERYTHING IN BETWEEN

LAURA LINN KNIGHT

To my children, Oliver and Grace.

May you always empower yourselves and others through knowledge, action, and curiosity.

***

Published by:
ULYSSES PRESS
PO Box 3440
Berkeley, CA 94703
www.ulyssespress.com

ISBN: 978-1-64604-614-0
Library of Congress Control Number: 2023943913

Printed in the United States
10 9 8 7 6 5 4 3 2 1

Acquisitions editor: Kierra Sondereker
Managing editor: Claire Chun
Editor: Scott Calamar
Proofreader: Sherian Brown
Front cover design: Amy King
Interior design: Jake Flaherty Design
Artwork: cover homes © Kate Kreker/shutterstock.com; interior home pattern
   © Iliveinoctober/shutterstock.com
Layout: Winnie Liu

CONTENTS

INTRODUCTION

My daughter stood in my room screaming in panic. I'd never seen her so completely terrified, and I couldn't understand why this was suddenly happening.

I wrapped her in my arms and tried my best to soothe her little body and mind.

Why was this happening?

I am a parenting educator, author, mindfulness and meditation guide, and former elementary school teacher. I have worked hard to help keep a calm and regulated home for my own family and help other families do the same.

There I was, facing down unprecedented fear in my then five-year-old, and I was baffled, scared, and beyond worried.

I stayed up all night, using countless tools I knew to help her calm herself. That night was the start of several months of intense anxiety for her.[1] During that time, I used every mindfulness tool I had and even brought in outside help to try and support her.

She was completely overwhelmed with worry, and nothing we tried helped her to fully regulate. I immediately started consulting with doctors. I put on my research cap and started making phone call after phone call. I talked to every specialist I could find, but I felt as though I was getting nowhere.

My heart broke for my daughter, and I felt that awful mama pain of not being able to fix the horrible fear she was experiencing.

Then, through a series of doctors, I was introduced to a functional medicine doctor (more on what that means and where to find one in Chapter Six) who ordered an organic acids test (OAT) for my daughter.

The results?

My daughter had high levels of mold in her body. My son and I had also been tested, and we both had mold in our bodies as well. Later my husband would be tested, and his results would also come back positive for mold.

And thus began my journey into the world of mold: addressing the source of our family's mold exposure and the consequences on our physical and mental health, discovering my own history with mold when I fell ill as a teenager, and learning about the millions of people suffering around the world from a condition that I didn't even know existed.

From that moment, we began to heal my daughter.

I began to heal myself and the rest of my family as well.

As I shared our experiences with family, friends, the families I work with, and my blog readers, I started to receive an abundance of phone calls from others who thought they might have mold themselves—a child who was unable to regulate emotionally and was delayed in his learning, several people who thought they had multiple sclerosis only to find out it was a misdiagnosis, a friend whose son's legs were covered in rashes, a mom who was working with multiple doctors but unable to heal her health problem until she found mold in her home, and the list goes on.

In the last two years, I've connected with a vast network of families experiencing both physical and mental health challenges, and countless times, when I've suggested someone should have their home and/or their body tested for mold, the results have come back positive. Those experiences with this cross section of my community have been a wake-up call; mold is more common and insidious than I ever realized when I was first starting my own family's journey.

MOLD AND YOUR HEALTH

What if I told you that your many health problems may be due to mold?

What if you had a solution for your child's asthma and hyperactivity?

What if you didn't have to suffer from mental health concerns and chronic fatigue?

Would you be thrilled if those migraine headaches would go away? What about your sensitivities to foods?

With an estimated ten million Americans suffering from mold toxicity, it is an illness that is causing dysregulation, anxiety, mental health conditions, rashes, hives, and much more.[2]

Of course, I don't think that everyone has mold. Nor do I believe that all asthma, migraines, or other illnesses are always due to mold.

However, I have had enough of my own experience with mold, and helping others with their mold problems, that I now understand this silent and toxic fungi that is affecting millions upon millions of people.

Mold often goes unseen and undetected. Many are living with it every day, without knowing the physiological, physical, and mental health effects it is having in their lives.

It may start with a small tear under the window, or perhaps it came into your home before your new roof was installed. Or maybe the mold isn't in your home at all, but rather in the old office building you worked in a decade ago ...

Unfortunately, mold is more common than we think and more destructive than many know.

It is a silent but all-consuming health concern for the many people who are allergic to it. And even with all the scientific knowledge available to us in today's modern world, mold is rarely understood by most medical professionals, and too many people are needlessly suffering because of it.

I now realize I had my first experience with mold as a teenager, which caused horrible intestinal problems, migraines, infertility issues later in life, and anxiety.

As a young child, my son had endless doctors' appointments to try and cope with health challenges he was experiencing—asthma, allergies, dark circles under his eyes—but we could never pinpoint a reason for his health challenges.

Then, my daughter was affected by the mold in our home and misdiagnosed with PANS (pediatric acute-onset neuropsychiatric syndrome) disorder. (Luckily, we quickly found out it was mold and were able to stop all her health-related problems.)

The more I researched, and the more I shared our story, the more I realized that our family's experiences were not isolated or even uncommon.

As my knowledge grew, I found more people seeking me out because they thought perhaps their children's emotional regulation concerns, hyperactivity, and medical conditions were a result of mold. Then I began to hear from adults who were worried about themselves or a loved one who was suffering for no apparent reason.

I began to uncover the dark stories of mold and the significant role it played in people's lives.

The silver lining is that for each of those calls, the negative mental and physical effects of mold have drastically decreased once I was able to direct those individuals and families to the right resources to clear the mold from their bodies and their homes.

In truth, I have seen nothing less than miracles happen for myself and those in my communities by understanding and healing from mold.

I have witnessed daily panic attacks disappear, life-altering health diagnoses reversed, intrusive thoughts significantly decreased, rashes vanish, asthma drastically improve, and more.

As a parenting educator and former elementary school teacher, I have devoted my life to supporting families and giving them the tools and resources they need to heal. My goal in writing this book is to empower you with information that will help your family heal physically and emotionally.

THE INSIDE SCOOP ON MOLD

I imagine you are reading this book because either you or a loved one has mold in your home or in your body. Or perhaps you aren't sure if you have mold, and you are hoping to learn more about the signs and symptoms.

Whatever has brought you to this book, I hope I can make your journey through the uncharted territory of mold easier to navigate than the one I have experienced. My wish is that you will walk away with the ability to take action so that you and the people you care for can regain a sense of balance and well-being in life. I truly hope that you will not have to go through the financial and emotional burden that I have because for too long, mold was pretty much an unknown and un-talked-about subject.

In the last couple of decades, of course, there has been more recognition brought to the seriousness of mold and mold toxicity. However, I still find most of the literature about mold to be thick, confusing, and hard to follow at times.

Now, I am not a doctor, and I do not claim to be an expert on mold. I *am* an avid researcher, and I have helped diagnose other families and children with mold toxicity and helped them find a path of healing. I believe that if you have mold toxicity (or a disease that is caused by mold), you can heal as well. I believe that knowledge is power, and you deserve to know the facts about mold so that you can act as an advocate for yourself and your family.

That said, I want to be clear: You should consult with your doctor before taking any of the medications and supplements I share in this book. Just because a protocol is working for me or for one of the families in my case studies, do not assume that it will work for you. The information and experiences I share here are meant to empower you with background knowledge to help you ask informed questions and understand your doctor's recommendations.

This book is not meant to diagnose or cure you. This book is meant to offer you and your family encouragement and knowledge in what feels like a void of support in the healthcare industry today.

My purpose here is to set you on a path of healing.

If you have struggled for years with rashes, headaches, fatigue, shortness of breath, diarrhea, constipation, muscle and joint pain, chronic sinus symptoms, abdominal pain, asthma, allergies, short-term memory loss, intrusive thoughts, anxiety, depression, or other ailments, and no doctor seems to be able to give you the proper answers, than perhaps you will consider mold after reading this book.

In most cases, mold is not overly difficult to diagnose with proper testing, and you can get help to clear it from your body, but first, you must have the awareness to look for it. You have to be your own advocate when it comes to mold, because there aren't enough doctors who are knowledgeable in this field.

Before we get deep into discussions about the signs of mold toxicity, testing your home and body, and healing from mold exposure, there are a few things you need to know:

Not everyone is affected by mold, and people who are affected will not always have the same symptoms. Two people can live in the same home that has mold and neither may get sick or only one of them may get sick or both could become ill with the same or different symptoms.

Most of the medical research on mold suggests that about 25 percent of the population has a genetic predisposition that makes their bodies unable to detox from mold naturally, which makes mold exposure especially dangerous for them.[3] Because their bodies can't effectively clear mold on their own, mold toxins will accumulate in those people's bodies and make them sick. Even worse, because the defenses that they do have are inefficient, their systems try to overcome the problem by flooding the body with antigens, causing inflammation that makes it even harder for them to heal and compounds their symptoms.[4]

Some people will have an immediate and noticeable reaction to mold exposure. Individuals who are autoimmune compromised and have another illness can experience even more drastic health difficulties more quickly after exposure to mold. Others, however, may have a slower buildup in their bodies depending on how often they are exposed to mold, thus allowing mold toxins to accumulate over time.

Mold can fall into different categories, and the type of mold that you are exposed to will largely determine the health effects the mold has on your body. Some species of mold can cross over into one or more categories. The three main categories of mold are: allergenic, pathogenic, and toxigenic.[5]

For the purpose of this book, I will mainly be discussing molds that are toxigenic (as my experience and work is mostly with mold toxicity) but will highlight a few other mold species throughout this book as well.

The most common indoor molds are *Aspergillus penicillioides*, *Aspergillus versicolor*, *Chaetomium globosum*, *Stachybotrys chartarum* (more commonly known as "black mold"), and *Wallemia sebi*.

Mold toxicity isn't widely acknowledged, but there are many people who have it. According to Dr. Ann Shippy, author of *Mold Toxicity Workbook*, "Mold toxicity is mainly caused by mycotoxins, which are basically poisons produced by mold."[6] Mold toxicity can have major health consequences, and if you do have mold toxicity, the most important first step you can take is to remove yourself from the source of the mold and then begin working to clear the mold out of your body.

According to Neil Nathan, MD—author of the books *Toxic: Heal Your Body from Mold Toxicity, Lyme Disease, Multiple Chemical Sensitivities, and Chronic Environmental Illness* and *Mold and Mycotoxins: Current Evaluation and Treatment*—mold toxicity is far more common than most people know, and researchers who are knowledgeable in the field of mold toxicity estimate that millions of people are affected by mold but are unaware that mold is the source of their problems.[7]

Mold is not easy to expel from the body once you have been exposed to it unless you have the proper supplements and medications. We will take a more scientific look at this in later chapters, but for now, it is important to know that if you have been exposed to mold, you cannot assume that by simply removing yourself from the source you will therefore remove the mold from your body. Most people will need to use binders (please see Chapter Eight for an understanding of binders) and/or medicine to help detox the mold.

Mold is found outdoors, and that is normal. It is not the mold outside of our homes that we need to be so concerned with, but instead the mold that grows inside that is a health concern. We will discuss how mold grows in more detail later in the book. Furthermore, people can be allergic to mold but not have mold toxicity. As previously mentioned, because my family and I had mold toxicity, and the people I have supported, interviewed, and studied also have had mold toxicity, that will be the focus of this book.

Mold toxicity can also be called "mold illness" or known as a biotoxin illness called "chronic inflammatory response syndrome" (CIRS). For the purposes of this book, I will be using the term "mold toxicity" for cohesiveness, yet it is important to be aware of the proper terms attached to your illness for research purposes.

EXPOSING MOLD TO THE LIGHT

In my previous book, *Break Free from Reactive Parenting*, I helped families create a *calm* and *happy* home through effective parenting tools that are designed for both the parent and the child.

Since writing that book, I have expanded my educational work to support families in creating a *healthy* home environment.

I wrote *The Toxic Mold Solution* to tell a story—the story of mold and how it is needlessly causing pain, and the best way to HEAL from that pain.

Using my personal experience, well-researched facts, the expertise of colleagues in the medical field, case studies, and my training in a multitude of healing exercises that will help to rewire your brain, I have written a book for anyone who thinks they (or a loved one) are affected by mold.

It is time to put an end to the misunderstood trauma that mold casts upon people's lives and begin the healing journey.

No matter where you are in the world of discovering and discarding mold, this book will help you to live the healthy life you deserve and will empower you with the knowledge that should be available to us all.

I never chose to become a teacher in the field of mold (rather, you could say that mold chose me), but my journey has served as a support for many, and I believe it will help you to transform your life as well.

Let's no longer hide mold behind the walls or away from the light of day. The time has come to expose the true harmful nature of mold and get you and your family the life you deserve.

I'll emphasize again—I am not a doctor, nor am I a mold specialist; I am a mother who couldn't accept her daughter's suffering.

I had to think outside the box and search for answers to a problem that was taking over our home, both literally and figuratively.

I was pushed to learn about mold and mold toxicity, and in my learning, I found a huge void in the material written for people like me who needed to know how to heal from mold toxicity but didn't have the time or capacity to make the deep dive into all of the available research on mold.

It was out of desperation that I read, highlighted, and powered through scientific journals and books about mold. It wasn't easy, and in my quest for understanding, I couldn't help but share about mold on my blog and then write a book for the other families who needed easy to digest information about mold toxicity.

I felt called to give a voice to mold and those suffering with this condition.

I was inspired to write this book for several reasons:

* I want to use what I have learned to provide a guide for families (both kids and adults) who want to know where to get started in their research and their healing. This book isn't a definitive guide to mold toxicity. It's a resource for families who are starting their journey. I have seen many families get out of their moldy homes or clear the mold from their homes, which were silently making them sick, and set off on a path of recovery. This book is an opportunity to share that knowledge with even more people who need it.

* I felt called to write the book I needed twenty years ago when I was on the carousel of doctors' appointments with no answers to my own ongoing health problems. I needed it then, and I am writing it now for those who need it and those who will need it.

* I am writing this book for my daughter, who had mold toxicity at age five and was almost put onto the same carousel of misdiagnoses as I was.

* Having walked this path myself, I know that finding the tools for combating mold in our homes and healing from mold in our bodies is just one part of the equation. The journey is long, and to be really honest, it's sometimes very hard. Having self-care tools is so important; without them, the overwhelm can set in quickly. And because this is my area of expertise, I wanted to write a book that equipped families with tools to support them mentally, emotionally, and spiritually as they move forward.

* I am sharing my experiences as well as case studies to show the diversity of experiences with mold toxicity and of approaches to solving the problem. We all deserve access to information when

"mold toxicity is currently affecting, to some extent, up to 10 million Americans."[8]

Most of all, I am writing this book so that you will know you are not alone.

When I was in the thick of my mold research, it felt like my world was coming apart at the seams. My children were suffering, and I did not know how to help them. I was battling my own health challenges while also struggling to find answers and solutions. It was all too much.

I read so many books about mold during that time, and while they were all full of important research, they were also all very scientific and sometimes over my head.

What I needed was a book written by (and for) a mom like me. Someone who had been through this experience and who knew how scary and overwhelming and isolating it could be.

I needed someone who could share their knowledge and the tools and resources they'd used.

Most of all, I needed someone who could give me some comfort that I could get through this. I needed the book equivalent of a warm hug.

That book didn't exist when I needed it. So I knew I needed to write it.

If I can give you nothing else, I want you to know this: You are not alone. There are others out here fighting this same fight.

If your brain works in data and scientific research, I will give you what I know and share where you can go to learn more.

If reading the stories of other families helps you feel empowered to ask new questions and advocate for your own family, I'll walk you through our family's story as well as numerous case studies of other families who are facing mold head-on.

And if you just need some affirmations to help you see your way through and keep going, that's the very heart of this book.

I would never have chosen for our family to be affected by mold toxicity, but I am grateful for solutions I have found and my ability to share those with others.

You can also find more information and resources on my website dedicated to shining a light on mold toxicity, https://thetoxicmold-solution.com/, and parenting resources for your family at www.lauralinnknight.com.

HOW TO USE THIS BOOK

In Part One of this book, I share my own personal story around mold. You will learn about my daughter's experiences, which helped me connect the dots on my own first exposure to mold in high school, and then my second big exposure when my husband and I bought our home in 2009.

You'll also learn exactly what mold toxicity is and the physical and psychological symptoms associated with it.

In Part Two, I walk you through the process of having your home and body tested to find out if you have been exposed to mold.

I share my experience with different testing options, including why the first test on our home showed no evidence of mold, even though later we found out that there were parts of our home where every inch of drywall was covered in thirty years' worth of mold.

I also lay out some of the most common options for testing, some of the pros and cons of the current options on the market, and how to locate experts who will help you interpret those test results and use them to start the process of healing.

In Part Three, we'll dive into healing your body and mind, which may have been greatly dysregulated due to mold. You'll learn about the correlation between mold, histamines, mast cell activation syndrome, and more. Then I offer you support and tools that you can put into practice right away and use to detox your body from mold, things that will have lasting results in your healing process and lead you on a path to creating a home for healing.

The chapters build upon each other, and I suggest reading the book from beginning to end.

Woven throughout the chapters, you will find case studies from people who have also been exposed to mold. All the names in the case studies have been changed to protect the people sharing the pain, grief, trauma, and frustration that mold brought into their lives. Each case study then shares with you how the person eventually healed from mold, the protocols they followed (or are still on), and the lessons they have learned in their mold journey.

If you are familiar with my work, you know that I am a huge fan of teaching families tools that support their emotional well-being.

I love all the information and research that is available to us, but I also find there can be high levels of information overload and confusion about how to digest that information. My job is devoted to researching and creating solution-based plans and tools to support and empower you.

The last section of each chapter will be devoted to tools that can help support you (and your family) on this journey when fearful thinking arises. I share these tools because I believe they are key to the process of healing.

This process is taxing to the system. It can be anxiety provoking, financially challenging, time consuming, and energetically draining. If you are reading this book, chances are good you don't feel well, and you're also supporting a family who does not feel well. Just getting started can feel so daunting.

The good news is that we can find answers—but the answers don't always make the process feel any easier.

So I want to make sure that you have *tools* to help you care for yourself, to help you practice gratitude, to help you tap into that bigger spiritual picture and develop a sense of inner peace and confidence that you can handle whatever is ahead during the healing process.

Because you truly can.

These are all the teachings I wish teenage me had known. It is the information I so desperately needed when my daughter was sick and ridden with anxiety from mold.

And it's the encouragement I needed when I was so afraid that we'd never see our way through this struggle.

Because what I hope you'll take away from this book is that if you actually do have mold, it is treatable. You can heal your body. You can detoxify your home.

The hardest part is just developing an awareness of what steps to take.

The good news is that I'm here to help you get started.

This book is an opportunity to take a complicated topic (that I am still learning more about every day, ha!) and get you started on a path toward healing. If you have mold toxicity, you will be amazed by the benefits you can achieve as you begin to clear it from your home and body. You will feel so much better after learning the practices this book is chock-full of.

PART 1

THE INS AND OUTS OF MOLD

CHAPTER 1
GOT MOLD?

That night I described to you in the introduction was a terrifying tipping point for our family.

My daughter's newly developed worry, it turned out, was the canary in the coal mine. Had her symptoms not presented in such a palpable and distressing way, we might never have uncovered the mold that was making our entire family sick.

For me, that's a very scary realization.

To understand why that moment, of all the symptoms of mold toxicity I now understand we experienced, was the one that changed everything for our family, you have to know a bit more about my daughter's ongoing struggles with anxious feelings.

Her first episode of all-consuming fears came in the middle of what was a scary time in all of our lives—right at the start of the worldwide shutdown for COVID-19.

We were at home under strict quarantine. My daughter had recently become nervous about sleeping in her own room. This wasn't the first fear we'd had to help her develop tools to overcome. Over the course of several years, she'd experienced anxious feelings around several specific things.

When she was about two and a half, she'd developed a fear of storms. As a parenting educator who specializes in mindfulness, I had a huge toolkit of resources to support her. We'd learn about what happens with clouds and rain during a storm. We'd read about storms. We'd go out in a storm. We would use her breath to calm her body down.

We were able to use enough cognitive behavioral therapy (CBT) and mindfulness tools to get through each particular fear—role-playing, storytelling, breathing exercises.

And to that point, we'd met every new fear head-on, and they were always manageable. We'd always been able to help her through the fears, and it never escalated to a level where it felt all-consuming.

Until that night in 2020.

In the middle of the night, she'd stumbled into my bedroom, half-asleep at first. And then she suddenly woke fully with a loud, terrified scream.

She couldn't fully identify what was causing her so much terror, at least not with words. But the feeling she was experiencing was so clear. She felt out of control. Her body was so racked with the physiological effects of her fear, and she had no way of knowing if she'd ever feel better. Her mind was spinning with worries that she couldn't solve, and she couldn't be sure she'd ever find her way out of that cycle. And the more her heart raced and her body felt unregulated, the more her fear increased and the harder it became for her to calm down.

There was nothing I could do. No tool I could use to reassure her. I felt so powerless and scared as she paced back and forth. Her tears poured down her face, and her heart beat so horribly fast that I was terrified for her.

I walked over to her as her eyes widened with fear and wrapped her in my arms. I sang her one of her favorite songs and kept telling her she was safe. Eventually, we got her into bed, and I read to her until she fell asleep.

After that night, I quickly found that those days of mild anxiety were behind us …

Cue full-on fear. She had to do half days at school for a few weeks when things were at their worst. She worried that if I wasn't right by her side, her worries would overtake her again and she wouldn't be able to find her way back to calm without my support.

For months, though it felt like years, we tried to figure out what was wrong with her. *Why this sudden onset of extreme apprehension?* I thought. Was she subconsciously processing COVID-19, and it was coming out in this other way? What was I missing?

Thanks to my professional background, I had a lot of experience with and knowledge about child development and the inner workings of the brain. Yet this just didn't make sense. Nothing about it made sense.

Every day, I had to be right by her side, keeping her fear at bay. Every night, she would wake up terrified, and I would be right by her side, ready with a book to read her back to sleep.

We consulted pediatricians, naturopaths, child psychologists, psychiatrists, and homeopaths. I pulled from every area I could to see if it would help me find answers.

One pediatrician thought she had PANS disorder, which can include a sudden onset of neuropsychiatric symptoms like obsessive-compulsive behavior, middle-of-the-night wakings, food restrictions, and more.

PANS was the closest thing that made some sort of sense, but I just wasn't fully convinced. Her experience didn't seem to match up with the symptoms of PANS quite right.

I kept digging, kept researching, and kept looking for answers.

Through the child psychologist we hired, I connected with a functional medicine doctor. A functional medicine doctor looks for the root cause of disease and will often check for toxins in the body. (We'll talk more about functional medicine doctors later in the book, I promise!)

She suggested that we test my daughter for mold.

I had a moment's hesitation. See, we'd actually had our house tested for mold several years earlier (a story I'll share in Chapter Five), and nothing had come up. And the idea of testing your *body* for mold was a new concept to me (more on that in Chapter Six).

But I knew enough to know that it was possible to be exposed to mold anywhere, and when you're in a desperate situation, when no one is able to give you any solid answers, you're willing to try just about anything.

So we agreed to see what would come back on an organic acids test.

The results gave us our first glimpse at the root source of so many of her struggles. What my husband and I discovered was that our daughter had incredibly high levels of mold in her tiny little body.

It just so happened that around the time we discovered this mold in our daughter's body, we were scheduled to do a small kitchen renovation. So in spite of the fact that the tests we'd done years earlier had come back negative for mold, we decided to have our home evaluated again using a different kind of analysis. (We'll talk more about the different options for testing your home in Chapter Five.)

We had just found a short-term rental home nearby, where we'd planned to stay during the short renovation, when we got the results back. What we found were really, really high levels of mold in our home (especially high in our daughter's room). In fact, the specialist who tested our house, a well-respected expert on mold with several publications, told us he couldn't remember the last time he'd seen readings as high as ours.

When the test results came back, we brought in an environmental consultant, whose job it was to figure out where the mold was growing in our home.

He walked through our house—the home that we'd always seen as a safe haven for our family—wearing a full hazmat suit. It felt surreal.

Ordinarily you hire an environmental consultant to find the locations of mold growth in your home. With their specialized tools, they're able to identify mold growth without having to physically see it, because it's obviously not practical to rip down every wall in your home looking for mold.

Because we already had contractors working on our house, though, each time he pointed out a potential problem area, I told them to open up the wall so that we could be sure. I was determined to find it all.

If there were any lingering doubts in my mind (or my husband's mind), if I'd had any question that the tests were set up to convince us to go through the intensive process of mold remediation, they vanished once we saw what was behind the walls of our home.

Every piece of drywall that came down was covered in mold. In the laundry room. In our playroom. In our living room. In our bedroom.

The test results had found especially high levels in our daughter's room, and when we took down the drywall in her closet (which shared a wall with the roof), it became clear why. The outside wall had NO flashing on it.

Flashing is a thin layer of metal that is used to prevent water from entering openings in the roof. Because no flashing was properly installed, there was a tremendous amount of water that had been leaking onto the drywall of her closet and creating the perfect habitat for mold to thrive.

We found mold behind almost every piece of Sheetrock throughout the house, but it broke my heart to find so much mold in my daughter's room.

With the help of our functional medicine doctor and child psychiatrist, we started my daughter on the antihistamine hydroxyzine, which is known to reduce anxiety (and can help the effects of mold toxicity in the body!).

Almost immediately, her bouts of intense worry and fear stopped! Honestly, I can't overstate how rapid the change was in her; it felt like we could see a difference overnight. Now, did it stop forever? No. She still has moments of nervousness, but those moments are short and developmentally expected for her age. Today, she is stable emotionally and mentally.

After her anxiousness decreased so dramatically, we slowly started her on nystatin (a medication used to treat fungal infections, including mold). After that, she made further progress, like being able to get a restful night's sleep again. It felt like a miracle to see her go from constant worry to a child who was calm, regulated, and happy.

In fact, it was a miracle! A miracle that we so quickly found that the source causing much of her fear was mold and that we were able to treat her for it just as quickly.

I later spoke about her case to a chief neuro physician at a renowned hospital in Northern California. I told him that I didn't think my daughter had PANS but rather mold toxicity.

He agreed! In fact, he said that he had been seeing an increase in the number of cases where mold was affecting the functioning of the brain and overall well-being.

I cannot even begin to tell you how incredible and validating that experience was. Many Western doctors are skeptical about the widespread, debilitating nature of mold toxicity. Lots of allergy specialists see it as a purely environmental problem: get rid of the mold in your home and your physical symptoms will resolve (not always the case!).

Here was this well-respected, highly trained, experienced doctor telling me that he thought we were on the right track.

By that point, though, I knew that my family had been suffering the effects of mold for years—and in my case, for decades. Why had it taken so long for us to get to the root cause? And really, what's the big deal about mold? How can it possibly make you so sick?

The answer is both simple and heartbreakingly complicated: a lack of awareness. So here's where I share what I know with you in the hopes that the knowledge will empower you to be your family's advocate and champion.

WHERE MOLD THRIVES

For mold to grow, it requires a source of moisture, something to feed on, and the correct temperature.

In places where the climate never drops down to freezing temperatures, mold is able to grow year-round (bummer, I know).

In freezing cold weather, mold will at least go dormant and take a break from growing and spreading during those cold winter months.

But if you live in locations like I have recently (California and Arizona), the temperature never gets cold enough for the mold to take a break.

In fact, where I lived (and had my big exposure) in Northern California, mold had all the right conditions to thrive: the leaks that developed in our room from our stormy winters; an inside bathroom leak from a roll of toilet paper that was flushed by a source that shall rename nameless (ha!); improper metal flashing on the roof that protects the home from getting wet outside my daughter's bedroom closet, which connected onto the roof (if you only could have seen the amount of mold she had behind her closet wall!); an attic fan that was improperly installed and recirculating warm air without pushing it outside of our home (mold loves hot air); and then all the drywall in our home to eat.

It was the perfect recipe for mold's favorite food, and based on the tests we had done in our home, some of the mold had been living in our walls for *thirty years*! Yes, that's right … thirty whole years. Mold has a very long life before it dies, even when the leaks stop or some of the conditions change, it can keep on growing and releasing toxins that make us sick.

But don't think that just because it's Northern California that you can't have the same problem in another climate.

When we first told friends and family that we would be moving to Arizona in 2021, everyone said that we would have good luck leaving the land of mold.

That is not the case though. Mold grows in Arizona as well (think monsoon season, leaking AC systems, and temperatures that stay warmer all year long).

According to the CDC:

> ...molds are very common in buildings and homes. Mold will grow in places with a lot of moisture, such as around leaks in roofs, windows, or pipes, or where there has been flooding. Mold grows well on paper products, cardboard, ceiling tiles, and wood products. Mold can also grow in dust, paints, wallpaper, insulation, drywall, carpet, fabric, and upholstery.[9]

In fact, the largest study to date of moisture damage in US and Canadian homes, which surveyed almost 13,000 households across 24 cities, found that 36 percent had mold or mildew and 50 percent had water damage, dampness, or mold.[10]

You can have mold almost anywhere you live!

However, here is a list of the top ten states for mold growth according to HomeAdvisor:[11]

1. Texas
2. Florida
3. Oklahoma
4. South Carolina
5. Nevada
6. Arizona
7. California
8. South Dakota
9. Tennessee
10. Kansas

A study by Quest Diagnostics ranked the ten worst cities to live in if you have a mold allergy:[12]

1. Dallas
2. Riverside-San Bernardino

3. Phoenix
4. Los Angeles
5. Chicago
6. Minneapolis–St. Paul
7. Saint Louis
8. Denver
9. Kansas City
10. San Antonio

You might be wondering what makes these cities and states so susceptible to mold and difficult to live in if you have a mold allergy?

Most of the time, you find that the places on these lists are more prone to natural disasters that cause water and/or property damage, e.g., hurricanes, tornadoes, and floods, all of which bring water into the home.

These cities and states also tend to have climates that stay warmer, and, as I stated above, mold loves heat. Furthermore, humidity brings higher levels of mold (hence why states like Florida and South Carolina rank so high on the list).

According to the American Academy of Allergy Asthma and Immunology, mold growth typically peaks during July in warmer states and October in colder states. However, mold spores can be found all year in the South and on the West Coast.[13]

Of course, you can have mold whether or not you live in one of these states, but it is helpful to know where mold tends to grow based on outside temperatures and environmental exposures.

Now that you have learned more about where mold likes to grow and you are aware that it could be in any home (including your own home), you may be wondering, "What's the big deal about mold anyway?"

WHAT IN THE WORLD IS MOLD TOXICITY?

To put it simply, mold toxicity is when you have been exposed to *mycotoxins*.

According to Dr. Neil Paulvin, toxic mold exposure is, in essence, mycotoxicosis. For reference, a mycotoxin is a toxic secondary metabolite produced by mold; in other words, mycotoxins are chemical waste released by mold, and those by-products have toxic effects on humans. Mycotoxicosis is used synonymously with mold toxicity.[14]

Paulvin says that mold illness occurs when a person does not have antibodies to mycotoxins, and mycotoxins have accumulated to high levels in their brain and body.

Neil Nathan says that "it is estimated by some experts that mold toxicity is currently affecting, to some extent, up to 10 million Americans. Most of those individuals are unaware of the existence of this condition ..."[15]

Now, just to be clear, not all molds will affect the brain and body so harshly. There are some types of molds that will not be toxic; however, because this book is for people who are having health problems, we will focus on the molds that do make us sick.

Some molds that are particularly toxic for us humans are:

Aspergillus. Aspergillus spores can be found in the air that all of us breathe; however, when it is inside the home and someone with a compromised immune system is breathing it in, it can cause a host of health issues.

Chaetomium. This mold is most commonly associated with water damage to the home. It loves dark places and eats drywall, wallpaper, baseboards, and carpets.

Fusarium. Fusarium is not as common, but it does leave its mark if you have it in your home. It also can result from water damage and likes to grow in damp areas in your home. Additionally, this mold has been known to be found in HVAC systems.

Penicillium. Penicillium mold makes the antibiotic penicillin. You may have taken penicillin in the past and been very grateful for this medical resource. But Penicillium as a mold growing in your home is not ideal; it can cause allergic reactions, infections, and other health problems for those allergic to mold.

Stachybotrys. This is most often called "black mold," although this mold does not always appear to be black! It can also be green or brown in color.

These molds are dangerous, but there is some hope because they need the right conditions to grow inside a building (more on that in Chapter Four).

If you do have these molds growing in your home or in a place where you spend time (and you have an allergy to mold or a preexisting health condition), you are more likely to develop mold toxicity.

Looking back on my own journey with mold toxicity, I can see how I was suffering long before I could fully identify what was going on.

Health issues can be like that sometimes. There can be a slow progression that you might not notice in everyday life, but later, when your awareness has grown, you are able to look back and see that there had been underlying sickness and pain.

That has been my personal experience for myself and also with both of my children. As I look back in hindsight, I can see how my children were affected by mold as babies; unfortunately, I just didn't know anything about mold at that time.

In the case study below, I'm sharing John's story, which is a reminder to me of how lucky we were to have discovered my daughter's mold exposure before it had lifelong negative consequences for her. John's experience with mold was also similar to my own, with long-term health complications, which I'll share with you in the next chapter.

Case Study #1: Losing a Lung to Mold in College

John was first exposed to mold as a young boy after having kidney cancer. Many years later, he would find during his junior year of college that his childhood mold exposure had caused mold to grow in his lungs, which resulted in part of his right lung being removed.

John's is a story of someone who was already autoimmune compromised because of the cancer he had as a child.

After undergoing successful treatment for his cancer at five, John's family decided to celebrate by moving out of their apartment and buying a home with a backyard so that John could get the dog he had always wanted.

What his family didn't know about their new home, but would quickly discover, is that it was filled with *Aspergillus* and *Stachybotrys chartarum* (commonly called "black mold").

Within ten days of moving into the home, John's energy drastically decreased and soon he was unable to get out of bed. He stopped being able to attend school, and doctors were confused as to why he was suddenly so sick again. Perplexed, they thought John's cancer had come back.

Then, a large storm flooded John's living room two weeks after the family moved into their new home. The family started cleanup efforts right away, but they immediately uncovered mold growing in the house. Mold can set in quickly, within 24 to 48 hours, but the extent of the mold in John's house suggested that this was not the first time it had flooded.[16] What they discovered was that the property was a natural dumping ground for water flow in the neighborhood, and there wasn't enough proper drainage around the house to account for the amount of water that collected in their yard during big storms. With nowhere else to go, the water had been seeping into the house, probably over the course of years and long before John's family moved in.

Although John had only lived in that home for a short time, having such a vulnerable immune system made his symptoms drastic.

John suffered from:

- Headaches

- Extreme fatigue (to the point he was unable to get out of bed)

- Allergies

- Rapid development of severe asthma

- Anxiety

- Tics in his hands and legs that he would later come to realize began during that time in his life and has lasted into his twenties

John moved out of his home immediately after finding the mold but would have a long and difficult path of recovery. Additionally, his parents would undergo a long legal battle with the person who sold them the home, who was unwilling to take responsibility for the mold within the walls.

Although younger than me, John would still need medical support during a time in history (late 1990s and early 2000s) when mold toxicity was relatively unknown and misunderstood. In John's story, we see this misunderstanding both medically and legally.

Eventually, though, John would find a knowledgeable doctor who would give him shots of Xolair, an injected IgE blocker, to help decrease the histamines in his body (a connection I will address in Chapter Seven). He would have some relief from the pain that had plagued him for so long and get a scholarship to a prestigious college.

In college, John's health took a turn for the worse when a minor cold seemed to linger with a persistent cough for months, accompanied by pain and a growing sense of pressure in his chest when breathing.

One day as he walked across campus, John was overcome with pain in his lungs and experienced difficulty breathing.

The mold from his childhood home had gotten inside his lungs … though it took months to be detected. John first began to notice that walking between classes on campus became gradually more difficult; he found himself severely winded from even a five-minute walk.

Ultimately, John was hospitalized with a collapsed lung caused by the mold filling his pleural space (the area around the lung) with fluid, thereby squeezing his lung and preventing it from expanding.

Surgical removal of the mold, and the part of the lung where it was growing, became a necessity. Additionally, during this process, his IgE level was more than 1600 (over 200 is considered high).

After the surgery, his Xolair injections were quintupled in dose, and a brand new IgE blocker, Nucala, was added. It took four years for his IgE and histamine levels to return to normal.

John's timeline of healing after he moved out of the mold is as follows:

- 2001–2003: John's headaches/fatigue/asthma gradually improved over about two years after he moved out of his moldy home.
- 2008: Xolair was given, starting seven years after exposure during a major flare-up of asthma. Xolair is a medication that can be used to treat asthma as well as other health problems.
- 2016: Mold-in-lung operation.
- 2022: John is doing better and is healthier than he was as a child, however, his body is still autoimmune compromised.

I asked John what he would want to tell people about mold given what he knows now. He said:

> Every time you move into a new home, you need to check
> for mold. It may have affected me more than most due to
> my compromised immune system, but even a perfectly

healthy person can become severely ill from it. If you do encounter mold in your home, listen to your body. There can be many effects, and you should try to be aware of any unusual pains, shortness of breath, or fatigue that you feel. Finally, if you've had these effects in the past, listening to your body needs to be an ongoing negotiation. My story shows how mold can cause serious problems even over a decade later.

I also asked him to share what treatments he found most helpful in his process of healing. Here's what he told me:

> Aside from surgically removing the mold, what was most helpful were the injected IgE blockers Xolair and Nucala. These are major treatments. One metaphor a doctor once used is that if an inhaler is a pistol, then each of these is a tank. With that said, if you have a severe case, you should ask your doctor if they may help.

Given the severity of John's health challenges and his ongoing need to be very cautious (especially with the COVID-19 pandemic), you would think John would be cynical and jaded by all the pain. However, I must tell you that in interviewing both John and his mom, I was taken aback by their positivity and gratitude for life.

In John's sickness, he learned the true gift of what it means to be able to wake up each day and have a chance to be present for whatever comes his way.

As we walk through the mess of mold together, I want to give you the opportunity to use this challenge as an opportunity for growth (just like John has). At the end of each chapter, I provide tools that will help you navigate your hardship with greater ease and mental well-being.

Let's dive right in with our first helpful tools!

CALMING TOOLS FOR SUPPORT

In this chapter, I want to share a bit about mindfulness and the power of having a mindfulness practice during your day.

For those of you who are new to mindfulness, I will share a sweet definition that my son said to me when he was four years old, "Mindfulness is just doing what you are doing."

I love the simplicity of that definition!

A more sophisticated definition of mindfulness that I use in my workshops and teachings comes from Jon Kabat-Zinn (professor emeritus of medicine at UMass Medical School, author of numerous books, and founder of the Mindfulness-Based Stress Reduction Clinic). Kabat-Zinn says, "Mindfulness—paying attention to something, in a particular way, on purpose, in the present moment, non-judgmentally."[17]

Both definitions remind us about the importance of being present and observing what is happening without judgment!

The wonderful thing about cultivating a mindfulness practice is that it can help you have less reactivity and be more present. Mindfulness is a great tool to add to your toolbox, and the following activities will benefit you as you recover from mold toxicity.

Mindfulness also ties directly into one of the fundamental tools that we are learning in this book—creating a calm home, mind, and body.

This is much easier said than done at times, especially when you are feeling so sick that you don't want to get out of bed.

That is where a mindfulness practice comes in for us, because with mindfulness we cultivate a calmer internal state, no matter what we are faced with. If you are too sick to get out of bed, then you can observe your current state without judging or shaming yourself. You can return to your breath and observe what is around you in the room.

There is growing evidence that mindfulness can:

- Reduce stress and anxiety
- Promote better sleep
- Sharpen focus and concentration
- Reduce obsessive thoughts
- Reduce depression
- Increase empathy and compassion
- Increase focus
- And more

That's why I believe it's so important to incorporate these mindfulness tools into each step of your journey to overcome the mess of mold.

TOOL #1: SELF-CARE CARDS

When working with fear (illness in the family system often brings fear), we must understand the brain. Understanding the brain will help you bring awareness to your own feelings and help you explain to your child what is going on when they are worried themselves.

Here is what is important to know about the brain. I will explain it simplistically so that you can share this information with your child:

The *upper part of your brain* is home to your prefrontal cortex. This is a wise part of the mind, and it can help you calm down when you are having big feelings.

In the *lower part of your brain*, you have what we call the amygdala and the brain stem. This area of your brain is working when you are angry, want to run away, or get super scared.

Helping yourself and your child label what part of the brain is activated when you are in fear or overwhelmed by the weight of sickness can help you move from the lower part of the brain to the upper part of the brain.

There are many tools that you can pair with noticing what part of the brain you are in, such as breathing, mindfulness, meditation, and a variety of somatic-based tools.

I suggest downloading my Calm Cards that I created for both children and adults from my website: www.lauralinnknight.com. These tools will really help you regulate and decrease your and your child's stress.

**SELF-CARE CARD:
10-SECOND COUNTDOWN
+ LIST WHAT YOU SEE, HEAR, SMELL**

If you're feeling overwhelmed with emotions or feel like the weight of the world is on your shoulders, do a 10 second countdown and list what you see, hear, or smell with each count.

Here's an example:

10. I see a large, lush tree outside my home.

9. I hear dogs barking and cars rushing by outside.

8. I smell a scent of food cooking in the kitchen.

7. I see pictures of me and my family on the wall.

6. I hear my heartbeat slowing down to a steady pace.

5. I smell the familiar candle in my room.

4. I see a book I want to read on my nightstand.

3. I hear my radio playing in the bedroom.

2. I smell the leftover rain from today's earlier showers.

1. I see my fuzzy slippers next to the bed.

WWW.LAURALINNKNIGHT.COM

I also created a set of Self-Care Cards to support you through the day, which are designed specifically for people who are healing. I've shared the first one here, but you can download the full set at https://thetoxicmoldsolution.com.

When using either the Calm Cards or the Self-Care Cards, try to label the part of your mind that is speaking the loudest. As you use the cards to regulate your nervous system, notice if you can hear the wiser part of your mind beginning to speak louder than your fear!

Sometimes, when you're feeling heavy emotions, you can take a pause to see your surroundings, do a small ten-second countdown, and feel calm enough to process those feelings.

TOOL #2: OBSERVING BODY AND MIND

For this tool, I would like to highlight the power of observation because it will benefit you greatly as this part of your journey.

Imagine for a moment that you are a mountain. The mountain stands tall and watches as the sun shines on its trees, the snow falls on its jagged rocks, the wind races wildly through the meadows, and the rain sprinkles through the trees.

As the seasons change, the mountain is observing each of these changes. The mountain is aware but not judgmental. There is an awareness accompanied by an acceptance.

For us humans, it would not be reasonable to be just like the mountain and have no feelings. We have the gift (and sometimes what feels like the curse) of being able to process our own experiences.

However, sometimes we are quick to label events and situations as "good" or "bad." We can get so caught up in the thinking mind that we forget to simply stand back from time to time and observe.

When you are dealing with mold, there will be many feelings, questions, concerns, and overwhelming experiences that will arise (and might have already come up).

It is imperative in those moments that you remember the mountain. You are watching your mind and body without judgment.

The gift you have (that the mountain does not) is to take that observation one step further and offer yourself compassion during the observation process.

Try placing your hand on your heart right now and silently or aloud say a kind phrase that is soothing to you.

Perhaps it is something like, "I know this is painful and hard. I am sorry you are going through this. I love you and things will get better."

It might feel funny the first time or two to do this but keep practicing this tool.

TOOL #3: COMPASSIONATE STATEMENTS

One of the tools I've found most helpful in my healing is the Buddhist practice of metta meditation. In metta meditation, we send compassionate statements to ourselves, those close to us, and out into the world.

Here let's focus on and expand upon sending compassionate statements to yourself.

First try to identify what it is that you really want for yourself.

For me, I truly want health, connected relationships with my family and friends, a safe place to live that is free of environmental toxins, to be helpful to others in my work, and to have a calm and happy mind.

In the space below, take a moment to write what you would like for yourself.

_____

_____

_____

_____

_____

Now that you have your list, can you shift those aspirations into compassionate statements?

Here is an example of what I mean:

May I have good health.

May I have connected relationships with my family and friends.

May I have a safe place to live that is free of environmental toxins.

May I be helpful to others in my work.

May I have a calm and happy mind.

Turn your writing into compassionate statements in the space below!

_____

_____

_____

_____

_____

I recommend you bring these three tools into your home as you go through your healing process.

CHAPTER 2
THE PHYSICAL SYMPTOMS OF MOLD IN THE BODY

My daughter's diagnosis was a turning point for our whole family, but especially for me, because as I learned more about mold toxicity in order to help her heal, I had a chance to reflect on health challenges that had plagued me for years—ones that only started to make sense in the context of what I was learning about mold.

So let me rewind here to tell you a bit about where I now understand my struggles with mold really began—in my teenage years.

It was the summer of Sublime. Not *Merriam-Webster's* definition of "sublime"—"something that is very beautiful or good; something that is extraordinary"—but the band Sublime.

The song "40 oz. to Freedom" played at high volume, at the apartment swimming pool across the street, and my friends and I didn't skip a beat as we dove and laughed our way from the hot cement into the deep, icy pool.

If you were around in the 1990s (especially in California), then you may already have some of the lyrics running through your head.

Sublime blended reggae, hip-hop, punk, ska, and all other types of music to make their own unique songs. Their music seemed to leap out of my CD player.

Those hot summer days with endless hours to hang with friends, listen to music, and shake with laughter until my sides felt like they were going to burst were magical. It was that magnificent gift of youth that seems to go by all too quickly. And yet lurking beneath those carefree moments was a stomach pain that started off slow and would eventually lead me to being nothing but skin and bones, curled up on more bathroom floors than I care to remember, crying in pain and a heavy dose of anxiety.

What I didn't know then, but I know now, is that the small two-bedroom apartment my mom and I had moved into during my eighth-grade year had mold growing behind my bed.

I wouldn't learn about that connection until almost twenty-five years later, when my daughter's symptoms stopped me in my tracks and forced me to find answers to what was plaguing her.

When the functional medicine doctor and I decided to pursue an organic acids test for my daughter, we also made the choice to test my son and me at the same time; after all, if my daughter was suffering from the effects of mold toxicity, it seemed logical to think that maybe we'd all been exposed.

Both of our tests also came back with high levels of mold in our bodies.

My mom had seen everything that was going on with my daughter, had been with me through the entire journey, and when I told her that I had mold in my body, she said something that shocked me: "You know, you had mold in your bedroom in the apartment we lived in."

She'd never mentioned it before.

She told me that she'd seen mold on the wall behind my bed, but she'd cleaned it and assumed that she'd done what was necessary to keep it from making me sick.

She couldn't have known that this isn't how mold works.

In the 1990s when we were living in that apartment, my mom didn't have the benefit of mold test kits that were easily available online and in most home improvement stores.

Even if she had, she wouldn't have known that if mold is growing on a visible surface in your house, there's almost definitely a source that you can't see.

And she'd had no resources to alert her to the fact that even if she could have completely remedied the mold growing behind that wall—no small task in an apartment, where tearing out drywall is usually frowned upon—the mold likely would have stayed in my body unless I was taking binders to help me detox.

That's because mold can bind to both fat and water. Most toxins that enter our body come up against our biological defense system, which runs all the way down to the cellular level. The membrane of each cell acts as a barrier, allowing some molecules to pass through while blocking others. But because mold can dissolve into both fat and water, it passes freely through any membrane in the body. For a majority of people, this won't cause major problems, because their bodies can effectively clear out those mycotoxins.

But, according to Nathan, in about 25 percent of people, the body lacks the gene necessary to eliminate mold. These people's bodies will try to eliminate the mycotoxins through the usual systems—by attacking the foreign invader with the autoimmune system and by binding it to bile and excreting it through bowel movements. But without that key gene, the process just creates chronic inflammation, and the body continually reabsorbs the mold. In those cases, binders often become necessary to break the cycle.[18]

It took me over two decades to learn about the effects of mold and properly clear it out of my body (and later my family's bodies, as you will learn more about in subsequent chapters). Over two decades of doctors, homeopaths, naturopaths, acupuncture, special diets, and tremendous amounts of pain ("tremendous" perhaps being an understatement).

Alas, as I returned from the swimming pool all that time ago, I lay on my bedroom floor. My daybed with the canopy, something I had so desperately wanted, was masking the mold I was unaware of.

Sublime was still singing in my head (the first verse of my favorite song, "Badfish," on continuous repeat). The words hold a different meaning in hindsight, almost as if they were an ominous warning of what was to come.

The lead singer of Sublime, Bradley Nowell, was sadly addicted to heroin and ultimately died from a heroin overdose. It has been said that "Badfish" was about his struggle with addiction.

Today, as I listen to that song, knowing the journey I was about to embark upon, I cannot relate to the heroin addiction, but the idea of mold "grabbing a hold of me," as Nowell says in his lyrics, speaks ample truth. A toxic fungus was creeping and crawling into my body and lowering my immune system. And, oh, how weak I would become. It was only a matter of time before I would be praying for a drastic healing.

MY PHYSICAL SYMPTOMS OF MOLD

I first noticed the stomachaches after an occasional dinner that felt too big or a slumber party where I'd consumed too many candies and slices of pizza. I'd had an allergy to dairy since birth, so the initial onset of abdominal pain wasn't completely out of the ordinary, as I had dealt with pain over the years any time I ate cheese or drank milk.

However, after a year or two in that apartment, my stomach pain seemed to be increasingly difficult to stand. It began to feel as though everything I ate disagreed with me, and my response was cramping intestines with shooting pain and then intense diarrhea.

There were times when I would be out of the house and then the cramping would begin. As if I had an internal timer in my mind, I knew that it was only a matter of minutes (or, if I was particularly lucky, hours) before the pain would swoop down like a hawk grabbing a mouse and there would be no escaping it.

At the onset of these moments, I would begin to make my way home to the safety of the small hallway and the one bathroom of the apartment that my mother and I lived in. It was my only hope for privacy as I mustered my way through what felt like an intestinal game of Twister.

Time went by, and I managed the pain enough to enjoy my first few years of high school, my family, my dear friends, and the joys of being a teenager. Of course, I had my moments of overwhelm and difficulty, but nothing compared to what was soon to come.

During my senior year of high school, someone very special to me suddenly passed away. She was one of those adults who always looked you in the eye and cared about what you had to say. Her smile was contagious, and her joy rubbed off on everyone around her. I was very close to her and her family, and we spent many days together after school and long weekends enjoying each other's company.

Then, one day, doctors discovered that she had a brain tumor. The tumor quickly began to take over, and despite everyone's best efforts to overcome the cancer, she passed away.

The sadness of losing her was confusing as a teenager, and my grief felt like it was a heavy blanket pressing upon my chest. I longed to see her one more time, to take away the pain of her death for myself and especially for her family. I wanted to hold her hand and see her sparkling eyes dance.

The grief equated to stress, and stress can significantly compromise the immune system. When the mind is stressed, the body becomes stressed as well, and that opens a door to increased infections, increased inflammation in the body, and increased levels of cortisol. (Cortisol is a primary stress hormone in the body and brings about a host of problems if it is being released at high levels over long periods of time.)

My body was now holding both mold and grief!

And it was mold's opportunity to take hold of my body even more than it already had.

My body started to shut down with headaches, and then the pain of trying to digest food made me fear eating rather than enjoying it.

I would eat and then immediately be doubled over in pain, lying on the bathroom floor, crying.

In an effort to understand my sickness, I went to my primary care doctor, who had me undergo many tests. I told him that the pain was so bad, it felt like I had something in my intestines trying to claw its way out.

They examined my intestines, shoved tubes down my throat, took ultrasounds and X-rays, but they couldn't find any visible culprit for my sickness.

At one of my last visits with this doctor, he sat my mother and me down in his office and told me that he was 99.9 percent certain I had Crohn's disease.

STRESS, GRIEF, AND THE MIND-BODY CONNECTION

When we're stressed, the immune system's ability to fight off antigens is reduced. That is why we are more susceptible to infections.[19]

Long-term activation of the stress-response system, and the overexposure to cortisol and other stress hormones that follows, can disrupt almost all your body's processes.[20]

A range of studies reveal the powerful effects grief can have on the body. Grief increases inflammation, which can worsen health problems you already have and cause new ones.[21]

Crohn's disease is a type of inflammatory bowel disease of the intestines. It can be a very serious disease, especially for someone as young as I was.

For the record, I *don't* have Crohn's disease. My doctor's diagnosis was proven false my freshman year of college when I had an emergency surgery for a cyst in my ovary and the surgeon was able to examine my intestines during the procedure and see that I had been misdiagnosed.

Nevertheless, in that cramped doctor's office as a seventeen-year-old high schooler, I was told that I would most likely die from such a serious disease and then sent out of the office to make my way to a graduation party.

The party was at the home of one of my best girlfriends. The counters were spread with platters of food, and teens raced from the second to the third floor of her gorgeous home, which opened to views of San Francisco and the bay. I couldn't enjoy the day, despite my best efforts. I kept pulling at my hair … It was beginning to fall out in clumps.

I called my father from the telephone in my friend's bedroom to tell him the news the doctor had given me, but I found it difficult to speak. My teenage mind was in shock.

My misdiagnosis is not an uncommon experience when it comes to mold. Although mold is becoming more widely accepted and understood, I have still helped many families in the past few years discover they have mold toxicity only *after* having been misdiagnosed and wrongly labeled with illnesses that they did not have.

What is even more difficult, and what was my experience, is that when the diagnosis is wrong, patients are often given medicine or put on special diets that make their mold situation worse.

After I was misdiagnosed with Crohn's disease, I was told to go on an "all-white diet." This type of diet consists of white bread, white chicken breast, white rice, white potatoes, white cream of wheat … You get the idea, right?

At its best, this type of diet lacks proper nutritional value. At its worst, this diet filled with gluten and foods that are higher in histamines is seriously damaging to someone in my situation. (We'll discuss all of this in more detail in Chapter Seven.)

That is just what it was: damaging. The idea of the "all-white diet" was that I would be able to better tolerate and digest foods without the constant stomach upset. What happened was that my body was flooded with foods that were hard for me to tolerate, and I quickly went from 130 pounds to 100 pounds on a good day.

My body was skin and bones to the point where I had to use a rubber mat to take a bath, just so I wouldn't bruise.

School became harder to attend—my energy was low, and my emotional vulnerability was high. With the support of my amazing mom, I was able to graduate from high school and was accepted into a college along the Central Coast of California.

I arrived sick and struggling, but I did arrive.

My mom sent me with an oversized bin of food in case I couldn't eat the food being served in the cafeteria. It was stuffed full of container after container of tuna fish and Ensure. *Yuck!*

My guy friends in the dorms nicknamed me Skeletor. I was a size zero, which honestly is just not the size my body is intended to be.

I sought out resources from the Disability Resource Center (DRC) at school for extra support. The DRC was nothing short of a miracle as they helped me with my poor health and the drastic increase of homework/responsibility.

Eventually, though, I found a diet that worked well for me (more details on that in Chapter Seven), and having gotten some distance from the apartment where I'd been exposed to mold as a teenager, I started to feel better.

In fact, my stomach healing felt so dramatic and complete that in 2007, I decided to stop taking the medication that had been prescribed by my doctor back when my stomach pain had gotten severe. At the time, I thought stopping the medicine all at once would be a good idea—like ripping off a Band-Aid.

Boy, was I wrong.

What I understand now is that the medicine I was given was helping to mask a lot of the side effects of mold exposure. Going off that medication, without support from expert medical guidance, quickly became one of the most excruciating experiences of my life, and honestly, given my health up to that point, that's saying something.

I felt like I wanted to crawl out of my skin. I was flooded with anxiety, bloated, and fatigued, and I was emotionally drained by the whole experience.

Luckily, thanks to a very steadfast support network, as well as the help of my family and my sweet future husband, I persevered through it, healing slowly over the course of years.

And then in 2009, my husband and I moved into the home in Northern California where we planned to start our family. (Yes, the same home I later discovered was full of mold that would make my family sick for years before we discovered the root cause.)

And suddenly it felt like all of the progress I'd made was slipping away.

In that house, I'd quickly start to notice an uptick in physical and psychological symptoms.

My anxiety started to ramp up almost immediately. I'm someone who is always looking for ways to take better care of myself, so I tried a number of different protocols to try to address my anxiety. No matter what I tried, though, I would just end up feeling worse.

Both of my pregnancies were difficult for me. I had a hard time conceiving my firstborn; and then, both times I was pregnant, I found myself with extreme morning sickness. I would vomit all day long (it should be called all-day sickness rather than morning sickness!) and had gestational diabetes with each pregnancy. I had ovarian cysts, experienced bouts of fatigue, and struggled with many food sensitivities.

Of course, it could be argued that all those symptoms would have occurred without the presence of mold. However, we must remember that mold toxicity lowers the immune system.

When our immune system is lowered, it makes illness, infection, and chronic health conditions come about more frequently.

In my case, even without understanding that mold toxicity was the cause of all of my suffering, I could feel the stress of its effects growing stronger in my body the longer we lived in that house.

With a deep desire to figure things out, I sought as much health and nutrition advice as I could from naturopaths, homeopaths, Ayurvedic specialists, doctors, and books, but it wasn't until my daughter's diagnosis that I could see the thread that connected all of these experiences—mold exposure.

PHYSICAL SIGNS OF MOLD

Now, let's zoom in on you.

Have you had persisting symptoms such as coughs or allergic reactions like itchy eyes, headaches, congestion, or shortness of breath that you can't seem to pinpoint?

Maybe you've been prescribed medication by your doctor to help alleviate physical symptoms you're experiencing regularly, but they're coming back over and over again.

Maybe you've already done your own research, tried different diets and foods, and experimented with everything you can think of to help get rid of the symptoms once and for all.

As you know, I have been there, done that.

The signs and symptoms of mold toxicity can be ambiguous. Like I said before, I'm not saying that you absolutely have it if you have multiple signs and symptoms. But if you've had health issues that haven't gone away—no matter what you tried or how many times you've just ignored them—take a look at the following list of physical effects that mold toxicity can have on the body:

- Eye irritation (can include watery eyes/tearing)
- Sneezing
- Coughing
- Light sensitivity
- Sore throat
- Skin rash
- Headache
- Lung irritation
- Increased frequency of asthma attacks
- Eczema
- Runny nose
- Congestion
- Dry skin
- Muscle cramping
- Aches and pains (including joint pain and morning stiffness)
- Diarrhea
- Brain fog
- Blurred vision
- Inflammation in the body
- Skin tingling and numbness
- Shortness of breath
- Vertigo and tremors
- Abdominal pain such as cramping, nausea, and bloating
- Irregular appetite
- Night sweats and/or temperature regulation issues

Another case study that I would like to share with you comes from a woman who spent years seeking professional advice for her own health concerns as well as the health concerns of her family.

In Athena's case study, you will see just how damaging mold can be physically (displayed in a picture of her son's serious leg rash) and the powerful healing that took place once Athena moved her family out of their home and began to treat everyone for mold.

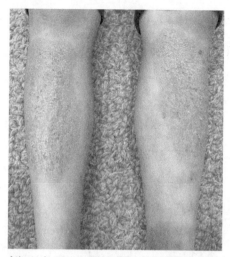

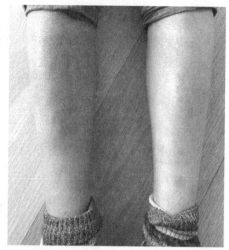

Athena's son suffered from a severe rash on his legs for three years while living in the home that had mold.

Athena's son's legs 6 months after moving out of their mold-infested home and starting multiple supplements to help him heal.

Case Study #2: Eczema, Tumors, Constant Wheezing, and MOLD

When Athena moved from San Francisco to Marin County, California, her son had always been sensitive in terms of allergies and gut issues. Because of his sensitivities, Athena's family didn't find it strange for their son to be having more intense seasonal allergies. After all, Athena's backyard housed over a dozen fruit trees and flowers!

Additionally, Athena's son had always had small issues with eczema, and when his eczema on his legs started appearing, they weren't alarmed (at first!). However, after a few years in the house, the eczema took over her son's shins, and no matter how clean their diet was (he had been on a restricted autoimmune diet all his life) and how many supplements he took, it wouldn't clear up the eczema.

During COVID, the family got an Australian labradoodle puppy, a breed known for being hypoallergenic, and Athena's son began having trouble breathing and had to be taken to the hospital. They gave up the puppy after two months, threw out all their rugs, and kept air purifiers on all day. Although their son's symptoms were not as acute, he was having a stuffy nose, red eyes, and sneezing in addition to his raw rashes. His pediatrician finally suggested they do a mold test to rule out the possibility of mold!

Athena has a beautiful home in Northern California. It is the type of home that many would love to live in, with its revered school district and the green mountains that surround the town.

Beneath the beauty, though, we yet again find mold...

Athena discovered the mold in her home in 2021. She used air-sample testing through a local mold specialist, and after finding mold in the crawlspace, she decided to get more extensive testing throughout the house: DNA, cultures, and pathways testing—testing developed by mold expert John Banta, which uses proteins to isolate specific spots around the house.

Athena had an inspector come after the test results and do a home inspection, and he found many areas of neglect and poorly patched work from the former owner, which pointed to a more systemic problem.

"We found that the water issues were more to do with the land and underground springs as well as being at the bottom of a hill and the target of runoff water from the houses up top," says Athena.

"Our family expressed symptoms [of mold] differently," says Athena. "I was in the house the most, so I had the most exposure. In three years, I had four surgeries for noncancerous tumors in my throat, uterus, stomach, and breasts. I had constant headaches, irritability, and aching joints as well having low white blood cell counts and anemia."

Her husband had consistent wheezing all year-round. Her daughter had recurring yeast infections, sleep troubles, and anger issues.

They began working with a naturopath who was a student of Dr. Neil Nathan, which meant she was experienced with mold issues. The naturopath started the kids on antihistamines and a detox program using activated charcoal and avoiding foods high in histamines. Athena and her husband did a similar program with the addition of nasal sprays and Medi-Clay-FX (calcium bentonite clay).

They moved out of their house, threw away all their soft furniture, stuffed animals, rugs, mattresses, and bedding. They are currently in the process of getting the mold cleaned out of their home so they can move back in.

I asked Athena what she would want to tell people about mold given what she knows now. Athena said:

> It's been such a life-changing experience to understand how much our environment impacts our health and how much of a difference we can make in our family's lives being more aware of this. It's so important to not ignore the things our body is telling us: allergies, rashes, headaches, and sore throats are indications that something isn't right around you.

I also asked Athena to share what has been most useful to her on her path of recovering from mold. This is what she shared:

> The knowledgeable people around us who have helped us understand what we were dealing with. The inspectors, doctors, and contractors have really helped us to make sense of our problem and help us find solutions; but mostly I am filled with gratitude for the families that have shared their mold stories with me, which has helped me figure out the more complicated and nuanced emotional issues with our recovery.

The physical symptoms of mold can be confusing. You may find yourself relating to many of the symptoms on the list. Or, like me, perhaps you only relate to a few.

We must also remember that not everyone is allergic to mold, and living in mold will affect everybody differently.

My own home was a perfect example. My husband's high levels of mold manifested physically as fatigue, IBS (irritable bowel syndrome), and cognition issues such as memory problems and headaches from time to time.

My daughter experienced mild asthma, eczema, and fear (more on that in the next chapter).

My son had asthma, eye irritation, bags under his eyes, sinus congestion, coughs, and a predisposition to catching colds and getting sick.

My symptoms included migraines, increased nausea, difficult pregnancies, IBS, fatigue, food sensitivities, some brain fog, and anxiety.

Having a checklist of physical symptoms is helpful in understanding how mold can manifest in the home and trying to determine whether you indeed have it.

If you are the slightest bit suspicious about mold at this point in the book, it is good for you to get a mold test for your home as well as for your body (more on that in Chapters Five and Six).

CALMING TOOLS FOR SUPPORT

As we wrap up this chapter, you might be feeling all different types of emotions. For some, you might be recognizing past symptoms and beginning to connect the dots between those and possible mold toxicity.

You may be confirming what you believed to be true: You most likely have been exposed to mold, and it could be the sneaky and silent culprit for your list of ailments. This may make you feel empowered to find the right doctor and start digging into next steps.

Or maybe you're still not sure if mold is at the root of the problem, but you are curious (more will be revealed as you continue to read).

Some of you might be feeling totally overwhelmed, and the thought of having mold scares you.

Whatever feelings you are experiencing at this point, I would like to invite you to take a moment to honor each of them. All feelings are welcome and valid in this space.

TOOL #1: SELF-COMPASSION EXERCISE

Imagine your friend has been sick for a long time now. She has been to many doctors, but none of them have been able to uncover why her illness isn't improving. She is frightened and sad and has come to you for support. Suppose that this is a dear friend of yours and you are in a stable place yourself. Think about how you would respond to this friend. What would you say to her?

How would you empathetically listen to her?

Now, place your hand on your heart, and listen to your own inner voice. Are you scared? Are you totally overwhelmed? What thoughtful and kind words can you offer yourself? What do you need to hear right now to calm your nervous system? From now on, I want to invite you to give yourself the same compassion you would give to a friend in need.

TOOL #2: BELLY BREATHS

This is an exercise that I teach children, but it works beautifully for adults as well. For a child, I ask them to find a stuffed animal and place it on their belly. If you have a stuffy laying around, feel free to grab it, otherwise the adult version of this exercise is to find a pillow that you can place on your stomach. Lying down on your bed or the floor, place that pillow on your stomach and watch it rise and fall as you take a deep inhale through your nose and exhale through your mouth. Continue to do this, and watch your body become calmer and more regulated.

TOOL #3: COMPASSIONATE BODY SCAN

The mind wants to latch onto the parts of the body that are in pain. It is hard to ignore a cramping stomach or a foggy brain.

The next time you have a few free minutes, I want to invite you to roll out a yoga mat on the floor or lie down in your bed.

Bring compassionate awareness down to the tips of your toes. You don't have to judge them or criticize them. If you have pain in them, send your breath into that space. Simply notice your toes and smile as you acknowledge their presence.

You will continue this same thoughtful and kind awareness practice as we move up the body—going from your toes to the soles of your feet and then the tops of your feet.

Notice your ankles, shins, knees, thighs, pelvis, hip bones, buttocks, abdomen, chest, arms, elbows, wrists, fingers, shoulders, neck, head, hair, and face.

Keep sending your awareness to each part of your body.

You may notice tingling, buzzing, pain, or temperature.

If there is a part of your body that feels extra achy, keep sending your breath into that space.

What can you say to the parts of you that feel icky? Can you talk to yourself with the same loving kindness that we practiced in Tool #1?

According to Mindfulness Body Scan Training by Jon Kabat-Zinn, "individuals that practice body scan meditation also report relaxing effects, especially individuals dealing with chronic stress or anxiety. Muscular tensions and tightness in the body often prevents you from feeling relaxed or calm in your surroundings. When you purposefully and mindfully relax the body, you feel a lot more at peace and relaxed."[22]

Scan the body and add some self-compassion for extra benefit!

CHAPTER 3
THE PSYCHOLOGICAL SYMPTOMS OF MOLD TOXICITY

Outside, it was a misty spring morning. A deer was behind the fence nibbling the plants she could reach with her mouth, and the chill of the day was beginning to fade into a soft warmth.

It was 2009, and I slowly climbed the steps to see my dear friend in bed—a bed that he seemed unable to leave.

It was as if he was paralyzed by his fear.

He had to be at work in ten minutes, but it was clear that he wasn't going to go. The blinds were drawn, and the room was almost too dark to see his face.

I didn't try to turn on the light. I knew that if I looked at him, I would see the pain in his eyes—his dilated pupils that told a story of panic and dread. I so badly wanted him to feel better.

"Can you let my work know that I am not going to come in today?" he asked.

I knew it wasn't just today. I knew he wouldn't be returning to work. Not for a long time. How could he? How could he work when his anxiety and depression had gotten so bad?

It would be ten years before he found out that his pain was from mold.

The psychological symptoms of mold are heavy, and these symptoms are so often unspoken, unknown, and pushed so deep inside the person who is suffering that they don't truly get the help they need.

In writing this book, the psychological symptoms of mold toxicity are what I felt most drawn to give a voice to.

I wanted to give them a voice because I saw them in my daughter, I experienced them myself, and I have worked with many dear friends and strangers who have also truly suffered in the dark without the help and support they need.

I wanted to bring them into the light because your mental health challenges may have a root cause that you can and need to address that has nothing to do with a genetic predisposition to mental health challenges.

But it starts with awareness, with knowing that something like mold toxicity can have a massive impact on your mental health. So let's dig a little deeper into the connections between mold and mental health.

HOW MOLD IMPACTS MENTAL HEALTH

We know that mold lowers the immune system, it can cause mast cell activation syndrome (more on that in Chapter Seven), it can take a serious toll on people physically, and it can overload your histamine bucket (which can cause you to feel fatigued, foggy, and more). That alone is a stress that is too much for most.

Additionally, it is quite common for people with mold toxicity to have anxiety, depression, intrusive thoughts, panic attacks, and other mental health illnesses that can be traced back to the mold in their bodies. I have heard many people with mold toxicity say that they feel like they are struggling beyond imagination.

The friend I described at the start of this chapter suffered from significant psychological effects within the first year of moving into his home. In fact, he had fallen into such a deep depression that he was unable to leave his bed and had to stop working for several months. He then spent several years with debilitating body dysmorphia that prevented him from going to many social functions or doing certain family activities.

The best psychiatrists were hired, but no SSRI (selective serotonin reuptake inhibitors) or alternative medication seemed to make a difference in his suffering.

His body dysmorphia became so pronounced that he would have panic attacks worrying what he should wear to work because he felt too uncomfortable with the way his body looked, and on hot summer days, going to the pool or beach no longer was an option.

His mental well-being was so impacted by mold that he dripped with sweat from anxiety, lay awake at night with a fearful racing mind, and in the worst of it, contemplated self-harm.

Eventually, he started taking a medication sometimes prescribed to treat severe panic attacks. This medication just so happens to have mast-cell stabilizing effects. It not only helps with anxiety, but it can actually benefit someone with mast cell activation syndrome (MCAS). MCAS, as it turns out, goes hand in hand with mold toxicity, and I'll talk you through that connection later in the book. It's interesting, therefore, that this medication made such a significant impact for this person, who unknowingly had mold toxicity.

Eventually, my friend had his home checked for mold after being encouraged by a loved one to test for it. He discovered that there were very high levels of mold in his home. The test also revealed that many of the molds in the home had been there for a long time, which explained why his symptoms were so intense upon moving into the residence.

The person then took an organic acids test and found mold in his body and has since begun a protocol of using antihistamines, charcoal, clay, and digestive enzymes to rid his body of mold (he also moved out of his home).

Does this person still struggle with anxiety, depression, and body dysmorphia? Yes, to a certain degree, he still does. However, it is not debilitating to the point that he is unable to work or function in the world. Furthermore, he is still at the beginning stages of clearing the mold from his body. So, although he is not actively living in a moldy home, the mold is still circulating within his body and needs to be cleansed further for the real benefits to show.

Unfortunately, this is not an uncommon scenario.

According to Mental Health Connecticut, "Mold toxicity can cause both physical and mental health problems and often goes undiagnosed, despite how prevalent it's believed to be. Finding this link could help millions of Americans understand what's really causing their symptoms and what they can do about it."[23]

The World Health Organization reports that one in four people suffer from a mental health disorder. While trauma and chemical imbalance are often the cause of these diagnoses, science is starting to uncover another culprit—mold.[24]

Psychological symptoms of mold can include:

- Anxiety
- Depression
- Insomnia
- Poor focus and memory
- Trouble retaining information and knowledge
- Panic attacks
- Confusion
- Disorientation

- Brain inflammation in the hippocampus (this is the area of the brain that controls memory, learning, and the sleep-wake cycle)
- Brain lesions in the gray and white matter, which impact the frontal cortex (the area of the brain that regulates impulse control, problem-solving, and social interactions)
- Increased sensitivity to pain and/or anxiety
- Irritability
- Brain fog[25]

Remember, mold disrupts the function of our organs, including the brain. A 2007 Brown University study of almost three thousand households found that participants who lived in homes affected by mold had a 34 to 44 percent higher risk of developing depression. But because mold exposure manifests differently in different people, there are a wide range of psychological symptoms.

I need to emphasize here, however, that even though there's a strong correlation between mold toxicity and mental health challenges, I don't think taking care of mold will always fix the problem. Addressing your mold exposure and treating your body is not the be-all and end-all.

But I do believe that if you're carrying mold in your body, healing from that is a very necessary step. When your body is so dysregulated, you fight an uphill battle to implement tools that will support your mental health. If your mold exposure is the root cause of anxiety or depression or emotional dysregulation, you'll need to detox to start the healing, but even if it's only a contributing factor, your body needs a chance to start regulating for you to get the full benefit of things like mindfulness practices, cognitive behavioral therapy, or eye movement desensitization and reprocessing (EMDR).

The following case study is another example of the devastating effect mold toxicity can have on mental health.

Case Study #3: Mold and Intrusive Thoughts

Tom's story illustrates the progression of mold toxicity and how it built up over time to affect his mental health.

Tom suffered from intrusive thoughts. Intrusive thoughts are associated with obsessive-compulsive disorder. People typically associate OCD with its physical manifestations (like extreme hand washing or counting repetitively), but intrusive thoughts are repetitive thoughts in the mind. Some people have coined this type of OCD as "Pure O," because it is obsessional inside the mind.[26]

Research has shown that intrusive thoughts tend to have five major themes:

1. Relationship (obsessing about whether you love your partner)
2. Sexual (this can be your own sexuality or the fear of sexual things)
3. Pedophilia (where the person obsesses about whether they could or would become a pedophile)
4. Harm (worrying about harming oneself or another)
5. Scrupulosity (moral or religious obsessions)

Tom is not the first person I know to have mold in his home and have corresponding intrusive thoughts, so this symptom is important to note.

Furthermore, I feel it is important for people to know about this type of OCD, because it is rarely talked about and is extremely stigmatized.

I got in touch with Tom through a friend when I heard he was suffering greatly from intrusive thoughts. I also knew he was living in an older apartment complex.

Tom's experience with mold began when he moved into his California apartment. He didn't know that he had mold in his apartment at the time (in fact, that would take years to discover), but he did notice quickly upon moving in that his mind began to fill with anxiety.

Tom's anxiety first began when he started teaching PE. Teaching was by far the most difficult thing he felt that he had ever done and the thing he cared about doing well more than anything else in life. He equated excelling as a teacher with being a good person, being strong, being a valuable part of the community.

In January 2011, Tom had his first experience with intrusive thoughts, which were distracting and sometimes felt debilitating; he worried constantly that they would impact his ability to be a good teacher and a good colleague.

Due to stress and ruminating thoughts, Tom took a year off from teaching. During that time, Tom lived out of state with his parents for a couple months, Yosemite for six weeks, and then traveled around South America for four months. Tom felt better, more relaxed, but not cured of the thoughts.

In July of 2018, Tom moved back into his apartment and started teaching again—this time in an indoor classroom instead of outside teaching PE. It was a drastic change from the year off. He was extremely overwhelmed and did not feel like he was doing a "good job."

During that time, Tom found that he was experiencing a new set of intrusive thoughts when he was at school. This is a very common experience; the specific nature of the thoughts may change over time, but the experience of obsessing and ruminating over these thoughts will be a common thread. Tom had a woman who worked in his classroom with him. He met weekly with her, and Tom says, "it was in these weekly check-ins with this other teacher that I first noticed my new, far more intense intrusive thoughts, centering around the possibility of me looking at her breasts."

Tom became terrified of looking at her chest. He locked onto her eyes and held there as a safe anchor.

Due to a change in positions at his school, Tom had a reprieve from his intrusive thoughts with the teacher. Unfortunately, they began again during parent/teacher conferences.

From January to March of 2020, Tom's thoughts became more intense, and the thoughts about looking at women's bodies and making them uncomfortable began to intrude everywhere.

Tom was finding it difficult to function in the world, and then the COVID-19 pandemic happened. To Tom's relief, he was able to teach remotely and interact less with the world.

Tom used that time to journal about his history, to read, and to seek support from a few close friends. Then, as I mentioned before, Tom and I had the blessing of connecting by way of a mutual friend.

I first heard about Tom's story over dinner with our mutual friend. The more I listened to the nature of his thoughts and the pain that he was in, the more convinced I was that he had mold in his home.

I told our friend to contact Tom immediately and have him test his home for mold. Unfortunately, I had just missed him as he had left for a long backpacking trip in Yosemite.

In November of 2021, Tom moved back into his apartment from Yosemite, and we connected about him having a potential mold problem. A month later, Tom discovered mold in the drywall in his bathroom by opening the walls himself (which he acknowledged in hindsight wasn't a good idea given he didn't have the proper protection for eliminating mold).

Nevertheless, Tom did get rid of the mold in his bathroom.

Tom then hired a professional mold inspector to come check his apartment for $275. The inspector didn't appear to be overly concerned with the mold that was still remaining, but he did say that while the walls were open in the bathroom, all the mold was free to disperse throughout the house, which was not good.

The inspector told Tom how to carefully remove the moldy gypsum boards with gloves and a mask, spray a disinfectant all around, replace the boards, and seal everything back up. He then told Tom to clean the whole bathroom and vacuum, focusing on all the corners where mold dust could settle.

The inspector also took an air sample from the outside and from Tom's bedroom, as well as a physical sample from a moldy piece of gypsum board in the bathroom. Each sample was $125, making Tom's total bill about $700 for the entire visit.

Tom says:

> When the results came back it was clear that there was toxic mold present in the bedroom air sample, as well as from the physical specimen from the bathroom. I called to ask about the severity of the amount they had found, and the inspector said that if I removed everything and covered back up the walls in the bathroom it would greatly help. He said the additional molds present in the air in the bedroom were not very high, so not very concerning, and I was the petri dish. If I was experiencing mold-related symptoms then I needed to go further, if not, I was done.

Tom recognizes the problem with using himself as the gauge for how much mold is present in the home and how toxic it is. He told me: "I acknowledge the flawed thinking that if the person is not presenting immediate symptoms then the home is clean from mold. Research shows that mold symptoms do not always present right away, and mold toxicity can take a long time to accumulate in the body."

Furthermore, Tom has since learned that air samples are not the most effective way to reliably measure amounts of mold in an indoor environment, as many of the most toxic molds are heavy and collect on the ground and in the corners.

Tom now believes that a more comprehensive and informative analysis of his situation would have cost closer to $1,000 to $1,500 (which felt expensive at the time), but Tom points out that:

> It may have given me either enough information to take more action, or have more peace of mind.

> In the end I spent over $1,000 dollars on what I thought was a cheaper way, piecemeal, and if I had it to do over, I would have done a bit more research and gone with a more reputable mold testing company.

After removing as much of the mold from his apartment as he could (and discovering that the apartment next to his is permanently closed due to what he suspects is mold or environmental toxins), Tom began to try to heal his body from mold.

In January of 2022, Tom began a three-month cleanse called the Bean Protocol, which was designed by Karen Hurd. Hurd's basic hypothesis is that toxins collect in the body, and while the liver attempts to detoxify, it stores the toxins in lipids. These lipids are reabsorbed as the liver bile travels through the intestines.

Hurd says that people don't rid themselves of these toxins, which could come from mold or a variety of other sources (this also includes hormones and neurotransmitters).

Hurd believes that soluble fiber is the best solution to this problem in that it binds to the toxins and is not reabsorbed by the intestines from the liver bile, and thus leaves the body as it's supposed to.

For three months Tom consumed no sugar, no caffeine, and no alcohol, and he strictly limited all starches. He also ate legumes with virtually every meal. Tom began to feel better and better throughout this diet!

Tom had gotten rid of the mold in his external environment and detoxified his internal environment. His intrusive thoughts "became less and less overwhelming day by day!"

Tom acknowledges that this wasn't an immediate fix (something we know to be true about detoxing the body), but he realizes that he is now doing things he would never have done before. Since his detoxification (and other healthy lifestyle choices like connections with friends and therapy), Tom has begun dating, and he is hanging out with friends and family in all types of situations. All these things are a big deal for someone who had debilitating intrusive thoughts!

Tom is a scientist at heart, and, for him, he believes that it wasn't just one thing that caused his intrusive thoughts, nor will it be just one thing that cures his intrusive thoughts.

He is grateful for the benefits he has received since clearing the mold out of his home. He also wants others who suffer from intrusive thoughts to keep up their therapeutic work, be honest with trustworthy friends, and to push themselves into the uncomfortable so that they can break through the barriers fear has created in the mind.

If you or a loved one are suffering from mental health concerns, you may want to investigate mold as an underlying cause. I know that mold was a major contributing factor in my own journey around anxiety.

You are not alone if you are experiencing any of these psychological symptoms. And, just like Tom in his story, there is so much to gain when you are honest with people who care about you and when you seek professional support.

CALMING TOOLS FOR SUPPORT

What did one slice of stale bread say to the other slice of stale bread?

"I used to hate mold, but it is growing on me!"

What did the sad mushroom say?

"Hang out with me! I'm a fungi!"

Okay, okay, mold jokes probably don't seem like the best choice in the book right now. But, truth be told, we have to find the moments when we can shine some light and love on this very painful topic. We have to find moments to smile (I hope those jokes at least brought a grin).

I do recognize the pain that you are going through. I know your pain because I have experienced the same pain, on a deep level. When our bodies and minds are affected by toxins, some days we just need a little extra positive encouragement.

Below are three more healing tools for you to use as you navigate this path.

TOOL #1: POSITIVE AFFIRMATIONS

Positive affirmations are positive statements that you can say internally or aloud to help combat negative thoughts. Positive affirmations tend to work best when you take ownership of them, so try to begin each statement with "I" or "My."

Examples of positive affirmations might be:

- I am becoming healthy.
- I am safe.
- I am trying my best.
- My body and mind are healing.

In the space below, write five positive affirmations that you can try out this week.

_____

_____

_____

_____

_____

TOOL #2: LOVING-KINDNESS MEDITATION

When I first began my mindfulness/meditation practice, I had a misconception about what it meant to observe one's thoughts. I believed that having equanimity around thoughts meant that I should watch my feelings and not respond to them.

This, of course, was a harmful understanding of my practice because it suppressed how I was feeling.

With a loving meditation practice, I learned how to observe my feelings while also offering kindness to myself.

When you practice this, you might notice your anger and place your hand on your heart while saying something like, "I know you're really angry right now. I'm so sorry your feelings are hurt. I love you."

When you are sad you might hold your arms around your body and say, "I love you. I know you are sad. This will pass, but until it does, I am here with you to support you."

Try talking to yourself in both the first person and third person to see what feels most comforting.

The second part of this practice is to send good thoughts, kindness, and love toward others during your silent meditation practice. (By the way, this is part of "metta," the Buddhist practice I shared earlier, which often uses a mantra or simple phrases.)

To do this practice, try sitting in a relaxed position in a chair or on the ground. Slow your breath with a deep inhale through your nose and a deep exhale through your nose or mouth. Try to let go of any worries (imagine placing them inside a cloud and sending each worry away through the sky). Try to notice your body and send your breath to the spaces where you want to further let go.

Your metta practice can begin with phrases toward yourself like: *May I be safe. May I be at ease and happy. May I be kind. May I be loved.*

You can then send that same loving kindness to one person who has loved and cared for you: *May you be safe. May you be at ease and happy. May you be kind. May you be loved.*

Notice if you feel warmth and love in your body as you say these words and settle into that awareness.

Then begin to send your loving kindness to your neighbors, community, friends, animals, strangers, and the world at large. You can also use this meditation with people you have difficulty with! Keep using the same phrases I have given or make up ones that feel comforting to you.

When we begin to cultivate this type of practice, we are better able to accept life as it is, one moment at a time, because our emotional well-being is rooted in love rather than despair.

As we build our toolbox, we find that those frustrating and anxiety-provoking moments with our health still arise, but we handle them with more grace.

TOOL #3: QUOTES

Sometimes I just need to hear a really good quote to put my life in perspective and find joy in the moment.

In my weekly parenting and well-being newsletter, I always include three quotes at the end for people to use for the week. You can sign up at www.lauralinnknight.com.

I have pulled some of those quotes for you here, and I encourage you to use any of these or to find quotes that speak to your heart.

Once you have found a few quotes that boost your mood, write them or print them and hang them somewhere in your home where you will notice them (the bathroom mirror is always a good spot!).

When we are discouraged (and mold is plenty discouraging, at times), we must find moments to combat that discouragement.

Five of my favorite quotes:

> "That is what compassion does. It challenges our assump-
> tions, our sense of self-limitation, worthlessness, of not

having a place in the world, our feelings of loneliness and estrangement. These are narrow, constrictive states of mind. As we develop compassion, our hearts open."

—Sharon Salzberg

"When I was a boy and I would see scary things in the news, my mother would say to me, 'Look for the helpers. You will always find people who are helping.' To this day, especially in times of 'disaster,' I remember my mother's words and I am always comforted by realizing that there are still so many helpers—so many caring people in this world."

—Fred Rogers

"You are loved just for being who you are, just for existing."

—Ram Dass

"Nothing can dim the light that shines from within."

—Maya Angelou

"Faith is taking the first step even when you don't see the whole staircase."

—Martin Luther King, Jr.

Do you have more quotes that you love? Write them in the space on the next page!

PART 2

TESTING FOR MOLD

CHAPTER 4

A Dream Home with a Dark Secret

On a sunny afternoon in 2007, at a time when I was again struggling with the effects of the mold exposure I'd experienced years before, I attended a talk where a small woman with dark-brown hair shared a thought that "the universe puts mountains in front of us because it believes in us."

I made that my mantra that day. I had been thrown into the fire already in many ways, but I knew it couldn't have all been in vain. I knew that it was going to be in life's most difficult moments that I would find strength and courage and allow my heart to open more to both myself and the world around me.

I believed then, as I do now, that every hardship is an opportunity for growth.

From that moment onward, the world around me began to soften a bit. I fell deeply in love with my boyfriend—who I eventually married—and was able to land a dream job as a teacher in the town where I grew up.

I'd screamed with delight when my acceptance letter arrived from Sonoma State University, announcing that I'd been accepted into their graduate program in education. With a heart full of anticipation for the future and sadness of leaving my sweet community in San Luis Obispo behind, I loaded up my mom's Volvo station wagon and set off for Northern California again.

I had been gluten free for a while at that point, removed from my first mold exposure, and was starting to get my footing back.

I was living in a small one-bedroom apartment with my pink-nosed rescue cat, Lady, discovering restorative moments for self-care when I wasn't busy planning lessons for my first grade classroom. I found support for my physical well-being in acupuncture, healthy eating, and establishing my meditation practice.

Eventually, even my health began to balance out.

My anxiety lingered, despite my deep desire that it would vanish in the way I felt it should, but I found solace by learning about spirituality, mindfulness, and somatic-based tools, as well as surrounding myself with other friends who had the same interests and goals as I did.

And then in 2009, I moved into a beautiful home that my husband and I bought, and I felt like life was starting to line up once more. I was over the moon to be living in what felt like my first real HOME.

In spite of what I know now—that our idyllic dream home was hiding a dark secret deep within its walls, that my family and I would ultimately absorb those toxins from that mold every day, day after day, for over a decade, that those toxins would affect each member of my family in slightly different ways but would make all of us sick— it's still hard for me to remember that time without thinking about how filled with joy and hope I was.

Not only was this the first house I'd owned, but it overlooked a preserved forest that was flourishing with redwood trees! Our backyard had an old oak tree whose branches swooped down toward the ground and a deck that spread along the back of the house, beckoning for you to come lie on it and watch the birds and squirrels.

I relished the hardwood floors, the vaulted ceiling in our master bedroom, and the quietness of our street.

In 2012, we brought my son home from the hospital to the nursery we'd lovingly decorated during what had been a difficult pregnancy.

I noticed right from the start that he had a tough time sleeping (hard not to notice when five minutes of sleep at a time felt like a good night, sigh!) and struggled with soothing himself. As a toddler he seemed to always have dark circles under his eyes; I'd find out later that these "shiners" were actually caused by allergies. The pressure from his stuffed-up sinuses caused the blood to back up in the veins around his nasal cavities, leaving him with deep purple shadows that looked like bruises.

As he grew, he started to develop asthma, and every time he caught a small cold, he'd develop what the doctors called "cough-variant asthma." He started out slowly with a dry cough and within a year had severe asthmatic coughs that would last all night long (it was so painful to watch his little chest laboring to fill with air and listen to his continuous cough for hours on end).

But I'm a researcher, and with my own history of food sensitivities, I figured that was a good first place to start. We started cutting things out of his diet and tracking what seemed to make a difference. But I didn't seem to be getting anywhere; nothing seemed to make a permanent difference for him.

So then we went to our pediatrician, who referred us to an allergy specialist. But in spite of all the testing we did, none of the Western doctors we consulted could give me any clear guidance on what we could do to help resolve his health challenges. In the end, the advice was almost always the same, "Eventually he should grow out of it."

As a mom, watching her child suffer, that answer just didn't work for me.

In my quest to help him through his asthma, I sought out every specialist I could find. We drove east to a holistic medicine doctor, north to a naturopath, west to a homeopath, and south to another naturopath (and then around all over again to Ayurvedic specialists).

We had some luck with homeopathy—treating his symptoms by leveraging the power of his body's own natural responses—but nothing completely solved his problems.

And then one of the naturopaths suggested that we have the house tested for mold.

All I could think was, "Really? No way. Not *my* house."

Everything I'd ever heard about mold to that point in my life told me that there was absolutely no way our beautiful home could have mold.

My house was always clean. There were never any musty smells; I definitely never saw any signs of mold growing on anything.

My house wasn't old—or at least, it wasn't old by California standards. It had been built in the 1960s.

We'd had a home inspection before we bought the house, and the inspector never raised any concerns.

And while I wasn't an expert on mold, I thought I knew enough to recognize whether my house was at risk of having mold. At one point I noticed a small leak, and I realized right away that I needed to open up the wall and dry it out. I even knew enough to look in to check if there was mold growing, but when I did, nothing raised any red flags. What I didn't understand was that if the mold was growing in the lumber, I wouldn't be able to see it.

Nevertheless, by the time the naturopath suggested a mold test on the house, I was also desperate for answers, so if this doctor thought it might help me to help my son, I was willing to try it.

I didn't really know anything about mold testing at the time my son was a toddler in 2014, so I hired the first company I found on Yelp to test our home.

The inspector who came out to the house explained that they'd collect an air sample in one of the downstairs rooms in our home. The sample would be sent off to a lab for analysis to determine if there were any mold spores present in the air.

Everything he said made sense at the time. If we were experiencing the effects of mold in our home, surely we were breathing it in. And if we were breathing it in, then it had to just be floating through the air in our home. Turns out that's not the case, but more on that soon!

When the results came back, I wasn't terribly surprised. He told me that we were fine. The lab had analyzed the air sample from our house and hadn't found evidence of mold.

To be honest, at that point in my life, the negative test was a blip on my radar. It was just one more failed attempt to find some answers to explain my son's asthma. At the time, I had no reason to question the test, so we moved on.

My daughter was born that same year. It had been another difficult pregnancy, but I was overjoyed to bring her home to her own cozy little nursery.

We noticed fairly early on that she was experiencing eczema, which is incredibly common in babies because of their delicate skin barrier. At first we weren't especially concerned about it. We were careful about what soaps and lotions we used and tried to treat it at home.

Unfortunately, nothing we tried seemed to make a difference, especially on the patches between her toes. When I talked to our pediatrician about it, they told me that asthma, allergies, and eczema are actually related conditions—all three conditions are linked to the body's inflammatory response. She gave me a prescription cream to treat it, which seemed to help.

And then when my daughter was about two and a half, we started to notice that she was having increasing fear, which culminated in that terrifying moment in our bedroom during the COVID lockdowns.

It was only through the experience of testing our bodies and finding mold on those organic acids tests that I learned how complicated the process for mold testing could truly be. Because when we had our home retested for mold in 2020 using an Environmental

Relative Moldiness Index (ERMI) test (more details about this testing option in Chapter Five), the readings on that test were clear and were confirmed as we opened the walls of our house and each piece of mold-covered drywall came down.

That was the beginning of a long and arduous process of fixing the problem. Knowing the extent of the mold in the house, there was no way we could move our family back into that toxic situation, and even if we'd decided to put the house on the market, we would have had to disclose the extensive mold growth to the new owners, and we wouldn't have wanted to subject anyone else to a health risk.

To us, there was only one option—make it right. What we'd originally thought would be a two-month renovation of our kitchen turned into a full renovation of the entire house.

It took well over a year, in part because we decided that if we needed to renovate the entire house, we were going to embrace it as a golden opportunity to change things and rebuild the house with a more modern feel.

In the end, though, we never moved back into that home. While we were trudging through renovations and living in a rental, we got an opportunity to move closer to our family members in Arizona.

So in 2022, when all of the mold remediation work was finally done, we sold our home in Northern California. We were able to document all of our work in the house and turn the keys over to the new owners knowing with 100 percent confidence that they were moving into a safe, clean home.

HOW I TESTED MY HOME FOR MOLD (THE SECOND TIME!)

When we had our house tested for mold the first time using an air test, the results came back clear, even though we now know the house was full of mold. Today I can look back and understand that there were likely several factors that led to a false reading.

Because we had a large house, testing only one room was probably never going to give us a full picture of what was happening throughout the house. It's possible that if we'd tested more rooms, we'd have gotten more definitive results.

The mold in our house was very old—probably at least thirty years old. So while that mold was still very much growing and putting off mycotoxins, there may not have been many spores in the air, and an air test relies on capturing spores.

The spores from certain heavier molds tend to fall to the ground, while most air tests recommend capturing the sample from three to six feet off the ground. You won't necessarily capture those heavier molds in the air.

We found the best results for our home at the time came from an ERMI test. We did this by having a mold company send an inspector to come conduct the test (but you can also do an ERMI test yourself for less money).

In my case, our ERMI test showed results that were much higher than one would expect to find in their home. For most of the rooms we tested, the findings were over the numerical value of 5 and reported that we had "high relative moldiness." For our ERMI test, a healthy value would be a Level 1, indicating that it is unlikely you have a mold problem. Further investigation was needed to determine the sources of mold. We also had some samples come back with "moderate relative moldiness."

NEW HOUSE, NEW TESTS

We had our best luck with the ERMI test when we were testing our home in Northern California. What I've learned since, though, is that not all tests are standard across the country, so I want to prepare you for that.

When we moved to Arizona, I was obviously very nervous that we'd move into another moldy situation. Closing on our house was a whirlwind—we only had ten days, not nearly enough time to get testing scheduled.

Even though it was a brand-new house, it could still have mold, although we could at least rest easy knowing that there weren't decades of mold growing behind the walls.

The best I could do was make sure we had a mold inspection before we moved in.

I found out, though, that in our area, no one uses the ERMI test, even though it was considered the gold standard in Northern California. None of the mold companies I called did ERMI testing! Our only option through a professional was an indoor air test coupled with a visual inspection.

Obviously after our experience with an air test back in 2014, I was hesitant to trust their results.

I found out that Lis Biotech offered the option for a DIY ERMI test. (A few companies offer this, but Lis is the company I'm most familiar with.) I'll be honest, though: the process felt overwhelming at the time.

So I had to make an imperfect decision. I opted to have a professional air test and visual inspection done first to see if there were any immediate red flags.

Both tests came back clear, but based on our previous experience and everything I'd learned about testing options, we decided to be cautiously optimistic but also keep an eye out for any physical changes in our family.

When my son's asthma started to flare, I held my breath, worrying that we'd had another false reading and that we'd actually moved into another moldy home.

An allergist reassured me that what we were seeing was more likely due to allergies to the local grasses, plants that didn't grow in

Northern California, and they started him on allergy shots to help him acclimate to these new allergens.

And while that made sense, I also couldn't shake the worry that there was something lurking in our house that was causing this uptick in symptoms.

Because we didn't have any other options through professionals in our area, I opted to do an at-home test from ImmunoLytics. The process was simple—we left out small trays designed to catch mold spores, then interpreted the results at home (you can also send the trays off to a lab for analysis). It was also affordable and had been recommended by several people in the field whose opinions I trusted.

I decided that if the test came back positive then I would spend the money and time figuring out the DIY ERMI test.

Those tests came back clear, which, coupled with the visual and air inspection, relieved my worries and made me confident that mold was not a concern in our new home. (That said, I'll continue to do an ImmunoLytics test yearly, and I do my own visual inspections regularly using the mold search checklist I share in the next chapter.)

I also asked our functional medicine doctor to retest our bodies (more on testing your body to come in Chapter Six). Although we all still had mold in our bodies, we didn't have a huge spike that would indicate that we'd had a new exposure in our Arizona home.

UNDERSTANDING YOUR TESTING OPTIONS

In the next two chapters, I want to share what I now know about the process of testing your home and your body for mold—all of the things I wish I knew in 2020, when my daughter's symptoms reached their tipping point and we found out that our house was covered in mold. Or in 2014, when our first home test came back negative while our family was suffering the effects of mold in our home. Or in 2012, when my son first started showing signs of mold toxicity. Or in 2009, when we first moved into that beautiful home.

In our case, had our family's organic acids tests (OAT) not come back showing signs of mold in our bodies, we might never have retested that house or learned what was at the root of our mental and physical health challenges.

I want to note, though, that the next two chapters about testing your home and your body for mold aren't in any particular order. Here's what I mean by that...

I wish I could tell you to get your house tested first and, based on those results, then to get your body tested.

Or to get your body tested first, and then use those results to determine whether to test your house.

Or that there was a single test that would give you all the answers you'd need to move forward toward healing your body and clearing your home.

I really wish that it was this cut-and-dried, but unfortunately, it's not.

You can have a lot of mold in your body and for some reason it will be suppressed and not show up on a test. You can have years' worth of mold buildup in your house but end up with a test that's not right for your particular situation and miss the mold in spite of a huge amount of growth, like we did.

So if you can afford to test both your home and your body, that's what I would encourage you to do, with the hope that one or the other will capture what's really going on.

I say that knowing this isn't an inexpensive process. In order to have your body tested, you'll likely need to be working with a functional medicine doctor or a naturopath (we'll circle back to talk more about naturopaths later as well) who can interpret the results and get you on a protocol that will work for your unique situation.

Professional home testing can be expensive too, depending on which one you choose, although, thankfully, we're now seeing an influx of user-friendly, effective home testing options.

Please do not let that become a barrier to finding answers. If all you can afford at this point are home test kits from your local big-box home improvement store, start there. At the very least they'll hopefully give you enough information to decide which steps to take next.

What I'm offering in the next two chapters is just an overview of the different testing options so you can make informed decisions as you move forward. I've also included case studies from two families that highlight the different experiences and outcomes that go along with various methods of testing so you can see how other families have approached the challenge.

The truth is that the process for testing is imperfect and slippery and sometimes overwhelming and frustrating. So, of course, I'm also offering some calming tools for support because self-care and self-compassion will be especially important in helping you navigate this uncertainty.

What I can promise you, though, is that knowing is always better than not knowing. Even if the process is tough, even if the answers are unexpected or upsetting, being armed with knowledge and awareness will empower you to make the right decisions for your health and the health of your family.

CALMING TOOLS FOR SUPPORT

When I write about our experiences discovering mold in our home, I can feel lots of emotions bubbling up. I feel sadness that our family suffered as long as we did. I feel pride that through tons of research and self-advocacy I was able to find answers and solutions. And I remember the feelings that came along with these experiences—fear, heartbreak, disappointment, hope, resolve.

If anything you've read in this book so far resonates with you—if anything seems familiar to you—reading these stories has likely brought up many emotions for you as well.

So I want to pause to share two visualization tools to help you process those feelings as they come up but also to maintain a sense of inner peace, compassion, focus, and optimism for the future.

TOOL #1: VISUALIZATION MEDITATION

Mold can leave us especially vulnerable to fatigue and stress, and sometimes I find myself feeling overwhelmed and unfocused.

As I go through my home my eyes take in the light, my mind notices the cluttered dishes that need to be put away, my body tenses when my two children begin to argue.

In the grocery store, I start to think about all the things that I forgot to add to my shopping list, and I am being pulled to purchase all sorts of new healthy snacks and scrumptious treats.

Of course, I do my best to bring mindfulness into each of these situations (check out my blog for lots of mindfulness tips and tools), but even when my mindfulness practice is strong, I find that visualization restores my nervous system each day. Visualization can help us prepare for those potentially overwhelming moments by letting us take a dry run first.

There are many ways to practice visualization meditation. You may find yourself visualizing a specific goal you have for yourself. For example, many athletes will take time each day to visualize themselves winning a big game.

You may also visualize how you want your day to go with your children—focusing on how you will respond, connect, and spend time with them for that day.

Or you may visualize your health for the day, seeing yourself being able to go for a short walk and eat healthy foods.

These types of visualization encourage you to picture how something will happen before it occurs. This can be a wonderful tool for keeping calm and optimistic. It can also help you set yourself up for greater success.

Another form of visualization that I absolutely love is nature-based visualization. In nature-based visualization you bring an image of the woods, a sunset, a flower, or anything from our natural world into your mind.

Sit in a relaxed position. You might sit with your back tall in a chair, or you might choose to sit crisscross on the ground. Imagine something in nature that inspires you, invokes great joy, and/or creates a sense of internal calmness.

As you summon that image, notice how your body begins to release stress and tension. You may also notice how your breath becomes softer and deeper as you settle into that image.

TOOL # 2: POSITIVE VISUALIZATION BOARD

Grab a poster or a big piece of construction paper, markers, old magazines, printouts of good health you have downloaded from the internet, an old book that you don't mind cutting illustrations from, and let's get to work (and by work, I mean let's start to have some FUN).

Making this board is your opportunity to put down everything you vision for yourself in the future.

Would you like a clean home that is mold free? Draw or cut out an image of a home and put it on your board.

Would you like to eat healthier foods and feel better? Add that, too!

What about being able to take a yoga class or go on a hike? Place those images on your board as well.

This is an opportunity for you to dream big and to imagine just how good life is going to get as you continue to heal.

Your visualization board can also help you to clarify your goals (now you see them right in front of you!), increase your positivity, boost your self-confidence, and get your priorities down on paper.

Use your newfound tools of meditation and mindfulness as you make the board, tapping into your innermost goals.

After you have created your board, take time each day to visit it in your meditation practice and/or when you need a reminder of the good that is to come.

CHAPTER 5

DOES YOUR HOME HAVE MOLD?

I f your home is like mine was, with no mold visibly parading itself on the walls, can you know if your house, condo, or apartment has mold? How can you avoid the experience that I had?

The scary truth is that mold can grow in any house. All it takes is one small leak from the AC or roof. Even just normal wear and tear on a house can make it vulnerable to moisture or leaks.

All mold needs are three things: moisture, food, and heat.

Think about leaving a loaf of bread sitting out on your counter for a little too long. It molds because all the right conditions are present—there's moisture in the bag, the bread provides food, and your kitchen is cozy and warm. Suddenly the whole loaf is just infested.

That's true in our walls, too. The mold in our homes grows because it has all three things. The moisture might come from a leaky pipe, a flood, a dryer that doesn't vent properly, or a bathroom with a broken exhaust fan. Household mold often feeds on cellulose in things like rugs, wood, wet boxes, and drywall. The heat could come from your HVAC system or from warm weather. It doesn't take much for mold to take hold.

The good news, though, is if that's happened in your house, there are steps you can take to find it and eradicate it. In this chapter, I'm going to walk you through some of the testing options for your home and some steps you can take to prevent mold from having the chance to grow in your sanctuary.

CHECKING YOUR HOME FOR MOLD

The best way to check your home for mold is by using a test. There are so many tests available that it can make choosing one very difficult.

Because I don't have experience with every mold company and their tests, I can't cover every test available. However, I will share with you the tests that I have known to work well (and not so well).

Having read this far, you know that mold is often hidden behind drywall, and when it is, we can't visibly see it. Also, despite what much of the research says about being able to smell mold, that was not the case in our home. We cleaned our home weekly and never had a musty smell that anyone noticed. Although, if you do have a musty smell in your home, all the more reason to test for mold ASAP!

As you saw with our experiences, mold tests can be inaccurate depending on the type of mold spores you have in the air of your home. When doing an air test, you need to know that some molds will show up better than others. Additionally, how the air sample is taken will be important. The height of where the sample is collected will make a difference in getting a reading that is more accurate. This is tricky because mold spores can have different weights, and if you sample too high or too low, you may miss an accurate reading.

Other mold tests (like the Environmental Relative Moldiness Index, a.k.a.ERMI) will use accumulated dust in the home as a sampling method. If you don't wait long enough for dust to build up, you may not get an accurate reading.

Not sampling enough of the home can also be a problem when it comes to mold testing. In our 2014 mold test, we only sampled one room that we knew had a leak after we moved into our home. We didn't account for any previous leaks in the home before we had moved in and the fact that mold can stay alive for years. When doing a proper mold test, you will need to get enough samples so that you are essentially testing your whole home.

WHAT TYPE OF MOLD TESTING IS AVAILABLE?

The current mold tests available are either DIY kits or tests done by professional mold-testing companies.

Many of the kits that you buy to test for mold yourself will require taking an air or dust sample with a provided sampling tool and then sending it back to a lab that you purchased the kit from.

Professional companies might come in and collect dust samples, take an air sample from your HVAC system, or wipe samples from your walls. They will usually conduct an inspection as well, where a professional comes in to look for visible signs of water damage, indications of mold, or hidden mold patches that you may have missed (think behind the refrigerator or, in my case, behind the washing machine).

The prices for both professional and DIY testing options vary widely based on a number of factors—the credentials of the professional you hire, the area of the country where you live, the number of rooms you want tested, the size and age of your home, the lab you work with, and the test type you choose. I've included links in the resources guide that will help you price out a couple of DIY options, but I would encourage you to do your homework and to reach out and get estimates from professionals in your area as well.

Based on my research, I am sharing three types of mold testing that seem to be your best bets.

TEST #1: THE ERMI (ENVIRONMENTAL RELATIVE MOLDINESS INDEX) TEST

Developed by the Environmental Protection Agency, this test analyzes dust samples collected from around your home (typically by an inspector, but it can also be completed with at-home kits you purchase) and sending those samples to a lab.

The ERMI test helps measure and detect whether there is an unhealthy amount of mold in your home by comparing test results to the national database of acceptable mold levels. For my family, our mold discovery journey included the ERMI test, followed by the inspection.

ERMI testing looks for thirty-six species of mold commonly found indoors. The first group of molds is associated with water-damaged homes and buildings, while the second group includes the most common toxic molds found in all homes:

Group One:

- *Aspergillus flavus*
- *Aspergillus fumigatus*
- *Aspergillus niger*
- *Aspergillus ochraceus*
- *Aspergillus penicillioides*
- *Aspergillus restrictus*
- *Aspergillus sclerotiorum*
- *Aspergillus sydowii*
- *Aspergillus unguis*
- *Aspergillus versicolor*
- *Aureobasidium pullulans*
- *Chaetomium globosum*
- *Cladosporium sphaerospermum*
- *Eurotium* and *Aspergillus amstalodami*
- *Paecilomyces variotii*
- *Penicillium brevicompactum*
- *Penicillium corylophilum*
- *Penicillium crustosum*
- *Penicillium purpurogenum*
- *Penicillium spinulosum*
- *Penicillium variabile*
- *Scopulariopsis brevicaulis*
- *Scopulariopsis chartarum*
- *Stachybotrys chartarum*
- *Trichoderma viride*
- *Wallemia sebi*

Group Two:

- *Acremonium strictum*
- *Alternaria alternate*
- *Aspergillus ustus*
- *Cladosporium cladosporioides* 1
- *Cladosporium cladosporioides* 2
- *Cladosporium herbarum*
- *Epicoccum nigrum*
- *Mucor* and *Rhizopus* group
- *Penicillium chrysogenum*
- *Rhizopus stolonife*

According to Eurofins Built Environment Testing, "A simple algorithm is used to calculate a ratio of water damage–related species to common indoor molds and the resulting score is called the Environmental Relative Moldiness Index or ERMI." [27] The company you hire should then help you compare your home's score to what you'd typically expect to find in houses without water damage or visible mold on or behind the walls. That ratio allows them to determine whether your home has unhealthy levels of mold in it. If you do the ERMI test yourself, you can hire an environmental consultant or find a DIY kit that helps you understand your results.

Alternatively, you can look online for DIY ERMI tests, which I considered when we were testing our new home in Arizona. In the resource guide at the back of this book, I included a link to the Lis Biotech DIY ERMI test kit, which is the option I'm most familiar with.

I also recommend getting an outdoor sample because there is a natural amount of mold that occurs both outside and inside the home. By getting mold samples outside of your home, you can compare the amount of mold that's possibly inside your home and make sure that they are not drastically higher than what is naturally occurring outdoors.

TEST #2: THE HERTSMI TEST

The HERTSMI (Health Effects Roster of Type-Specific Formers of Mycotoxins and Inflammagens, also known as HERTSMI-2) test is a smaller, more budget-friendly (though limited) mold test. Like ERMI testing, it uses MSQPCR (mold-specific quantitative polymerase chain reaction) analysis to identify mold species and then compares the levels of these species to what's considered acceptable levels. The collection process is the same as what you'd use for ERMI testing.

The biggest difference is the number of mold species each test identifies, which is what accounts for the variation in pricing. HERTSMI analyzes dust samples and tests for the most five common molds (*Aspergillus penicillioides, Aspergillus versicolor, Chaetomium globosum, Stachybotrys chartarum, Wallemia sebi*) frequently found indoors, compared to thirty-six with an ERMI test. The lab analysis with ERMI testing is more rigorous, so I would recommend the ERMI over this test if you can get it done.

TEST #3: DIY GRAVITY PLATES, SWAB TESTING, AND BULK MATERIALS TESTING

If hiring a testing company is outside of your budget, or if you're having trouble finding a reliable expert in your community who offers testing, there are numerous options for DIY test kits. I mentioned the DIY ERMI and HERTSMI kits from Lis Biotech above, which are much less expensive than hiring a company, but you also have other options if that process feels overwhelming or if those are still cost prohibitive for you.

Many home improvement stores and labs now offer inexpensive options for testing your home yourself. These typically use a few different options for collecting samples—gravity plates, swabs, and bulk materials.

The benefit of DIY testing is that it's relatively inexpensive compared to professional testing; most kits are less than twenty dollars, though there are often additional costs to get a lab analysis of your results. If you're planning to send things off to the lab (more on that in a minute), make sure you're clear on the total costs. Some labs charge per sample; others have a flat fee for a set number of rooms.

Gravity plates are a kind of air test, but unlike the one we had done in our home, where an air pump collected spores in the air, gravity plates collect spores that are floating free, close to the ground, in the space. You put a small petri dish in the room you want to test; the plate contains a solution that traps the spores and produces a visible reaction as the mold grows.

After an hour, you collect up your plates, and then seal them up. Depending on the test you've purchased, you may have the option to analyze the results yourself. You store your samples in a warm, dark area, and in a few days, you'll be able to look at the dish and see if you've got mold colonies growing in the solution. Depending on how many colonies are growing in your sample dish (remember, having low levels of mold in your house is very normal), you'll be able to determine whether you need to consider mold remediation.

The standards and measurements for each kit will vary, so it's important to read the instructions that come with your particular kit. But just to give you one example, here are the guidelines ImmunoLytics includes with their gravity plates: zero to four colonies is a normal range for a home, five to eight colonies signal an elevated risk of mold-related health effects, and nine or more colonies represent a hazardous health environment.[28]

Most companies also give you the option to send the plates to their labs to be analyzed; they'll be able to test the mold and give you a report on which strain (or strains) you're dealing with. I'm most familiar with ImmunoLytics's gravity plate testing. If you send in your sample for testing, they also offer a phone consultation that allows you the chance to ask questions and get recommendations about what your next steps should be.

If you know (or suspect) you have mold growing because it's visible on surfaces, DIY swab tests allow you to collect a sample using long-handled swabs. (Picture the ones we've been using for at-home COVID testing.) You send that swab off to a lab and receive an analysis of what kinds of mold you collected. ImmunoLytics also offers swab-testing options.

Some labs also offer bulk material testing. For these tests, you send off a small piece of drywall, carpet, wood—something you believe has mold growth on it. Keep in mind, though, that when you cut into these materials, you're likely disrupting spores that have settled and releasing them back into the air, so make sure you protect yourself properly if this is the route you choose to go. Personally, this option makes me a bit nervous because of exposure risk, so I want you to know that it is an available test, but it wouldn't be my first choice.

No matter what type of mold test you decide to use, you can do a visual inspection in your home as well. I have included a mold search checklist to help you with your inspection.

MOLD SEARCH CHECKLIST

- ☐ Check under sinks, behind the washer and dryer, and around the water heater. Don't forget to look at the bolts and pipes for any leakage or stains (including rust stains), which can suggest an earlier leak.
- ☐ Check windows and screens to make sure everything is tight and water won't drip in during a storm. Also make sure that dust and dirt don't build up in the openings that some screens have for water drainage, which can cause the water to pool and run back into the window casing.
- ☐ Inspect ceilings and walls for any signs of water damage. This can include anything from cracks and peeling paint to sagging or soft spots.
- ☐ Look for warping in your floors, which could be a sign that you've got moisture under the floors or damage to your foundation. If you've got hardwoods or laminate floors, keep an eye out for staining, discoloration, or soft spots. If you've got carpets, check for unexplained damp spots or buckling. If you've got tile, look for visible signs of mold or cracks in your grout.
- ☐ Listen when your house is quiet for the sound of dripping or rushing water. Catching water leaking behind the walls before it has time to wreak havoc on the hard surfaces is the first line of defense.
- ☐ Check your AC filters for musty smells or black fuzzy spots.

AN EXPERT'S OPINION

In trying to understand the best way to test a home for mold, I reached out to the prestigious mold expert Bill Weber. Bill has a tremendous amount of experience working with mold, and he has helped countless people discover and discard mold from their homes.

For this book, Bill wanted me to share with you that:

> the best mold sampling strategy can vary from property to property and occupant to occupant. There are several strategies depending on available funding (budget), purpose(s) of the testing, and whether or not a third party might be responsible for the [suspected] mold growth.
>
> When the mold growth is visible and the source is known, surface sampling using MSQPCR analysis would be a good choice to determine areas of settled spores and fragments ... If the results indicate elevated levels of water damage indicator species, small particle cleaning should be performed outside of the structural remediation area(s) to return the areas to normal fungal ecology.
>
> When mold growth is suspected and cannot be readily observed, mold sampling can help indicate whether or not there are hidden reservoirs. ... In most cases, sampling with a combination of dust collection and analysis using MSQPCR and fungal culturing can provide a lot of good information regarding the fungal ecology of the indoor built environment.
>
> In addition to testing for mold, ... it is important to consider the need for a qualified indoor environmental professional to not only perform the mold sampling, but also to assess the building, living conditions, and other environmental factors.[29]

You can learn more about Bill Weber and his work from his website https://avelar.net/articles_publications/team/bill-weber.

FINDING THE MOLD

Once you test your home using a reliable testing method, you will discover whether it has mold. If it does, you will need to find it!

Step One was to test, and Step Two is to locate it. This is the point where you might hire an inspector from the same company that did the testing or someone else who specializes in finding and cleaning mold.

When you select a test for mold and then hire an inspector, it's important that you put on your detective hat for your home. You want to think of every leak, each potential point of entry that water could have used to get behind your walls, and anything and everything that would have given mold an opportunity to grow.

If the inspector doesn't know the right place to look or you don't test in the areas where the exposure is, you might miss the mold.

Our family didn't have much to miss in 2021, because we'd fortunately already moved out of our home and were in the process of a small renovation. Our ERMI test showed that almost every room in the house had high levels of mold, so when the inspector we hired came to find the mold, we were able to open up much of the drywall with our existing construction crew, who were already on site for the renovation.

This, of course, is not going to be most people's experience. Just know that a skilled professional inspector can locate areas of mold without having to tear down all of your walls. Do your research, ask questions, and don't give up. You are taking big steps to help yourself and your family!

SIGNS AND SYMPTOMS OF MOLD IN YOUR HOME

Many of the people I talk to about mold aren't sure if they should begin with an at-home test or by hiring a company.

The feedback I usually share is to make a checklist of the signs and symptoms of mold in the home as a starting point.

☐ Your home has a musty smell. (It won't always smell, but if it does, that is a great indication that you have mold.)

☐ Someone in your home is having physical symptoms such as rashes, sinus congestion, headaches, coughing, sore throat, body aches, shortness of breath, and a host of other symptoms that have continued to progress and don't seem to have a medical solution. (I will share more about physical symptoms of mold in the next chapter.)

☐ Someone in your home is having mental health symptoms, such as depression, anxiety, intrusive thoughts, mental fatigue, poor ability to focus, and more. These symptoms do not respond well to treatment and continue to get worse.

☐ You have had a known leak in your home.

☐ You have visible water damage.

☐ You have seen mold somewhere in your home.

☐ You have an older home that may have had water damage in the past. (Remember, mold can live for a long time once it enters your home.)

☐ You live in a climate described in the introduction to this book.

☐ Your gut is telling you to check your home for mold!

If you find yourself answering "Yes!" to even a few questions on the checklist, you will want to look into getting your home tested.

Our next case study is a great example of how starting with a relatively simple inexpensive at-home test can give you the initial insight you need to start a more in-depth investigation. Brielle began her mold discovery by using a mold home test kit, which came back positive. That positive result confirmed that she had mold in her home, leading her to bring a professional company. This can be a great option for families as well!

Case Study #4: A Moldy Home

Brielle's passion and work has always kept her heart close to nutrition and well-being. She is the type of mother who feeds her two children organic food and keeps her home stocked with nontoxic cleaners and products. She is thoughtful about the environment she exposes her children to and is a caretaker to her family through and through.

It was surprising then, when one of her two children continued to have health issues that he struggled with. Her older son especially, seemed to have difficulty with constant tummy pain and gut dysbiosis.

Furthermore, Brielle was having her own host of health problems that included undiagnosed abdominal pain, inconsistent bowel movements, brain fog, and irritability around her menstrual cycle. She also had seasonal allergies, sinus infections once or twice a year, and often suffered from nasal congestion, itchy eyes, and sneezing fits. She dealt with painful periods, which she thought were normal, and excessive bleeding. It took two years of ruling things out before she figured out the root cause of many of these conditions was mold.

She first lived with the pain for six months before going to her primary care doctor. Ultrasounds, blood work, and an MRI brought no answers. She was referred to a GI specialist who scheduled a colonoscopy for March 2020, and when it was canceled due to COVID, she looked into other diagnostic tools that could bring her more answers.

She found a thermography clinic that was able to take pictures, using a thermal camera, of her abdomen and small and large intestines, around the area she was experiencing the pain. It confirmed significant inflammation. That was the first time she felt her symptoms validated, and she was determined to get to the root cause of that inflammation.

She had a GI map (stool test), which uncovered a slew of bacterial overgrowth, small intestinal bacterial overgrowth, and leaky gut. This was another piece of the puzzle, but she still hadn't gotten to the root cause of her inflammation mold just yet. Brielle adjusted her already rather clean diet, trying gluten-free, dairy-free, and alcohol-free foods and beverages, only to feel no improvements.

She was put on a number of protocols to kill off the bad bacteria and improve her overall gut health. She saw some improvements in her abdominal pain and bowel movements, but new symptoms also began: itchy rashes covering her legs and torso (a high histamine response can be caused by the body overburdened with toxins, especially mold). She was exhausting her time and money, with very little progress, and still hadn't found the source of her issues.

She was referred to a functional nutritionist, who Brielle worked closely with for the next year and a half, but even then, no one suspected or tested for mold. There was no visible mold in her home, and she had no idea about the connection between water damage and mold growth and her health.

Brielle and I happen to be good friends, and there were many early morning hikes after we dropped the kids off at school when we would talk about health, nutrition, and mold (yes, if you are friends with me, you are probably talking about mold from time to time!).

Brielle felt lost in trying to navigate the emotional and physical health issues that her family was experiencing. Here is how she explains it:

> I was frustrated and exhausted, spending time and
> money to make strides for my health, while feeling like

I was getting worse. I wondered how I got here, why I had such a disrupted gut, and questioned if I would ever understand what brought me to this point.

Eventually, a turning point came after she had remodeled the front part of her home and she decided to order an at-home mold test kit.

In August 2020, Brielle unknowingly increased her and her family's exposure to mold in her home, where they had resided for ten years. A partial home remodel began the same week as wildfires in Northern California. This resulted in Brielle and her family being confined to her master bedroom as the primary living space, which she later identified had high levels of mold. They had one large air purifier, which ran 24/7, and later she was told by a mold consultant: "If you had not had air purifiers in your home, you and your family would be significantly more sick." She couldn't open the windows because the air quality was so poor from the wildfires. And they were sleeping, eating, and lounging in their master bedroom for three consecutive months.

The tipping point that led to Brielle asking her functional nutritionist to test her for mold in her body happened just two months after her remodel construction began. She went on a weekend getaway to Pismo Beach and returned home with a sinus infection confirmed by her primary physician.

Brielle notified her functional nutritionist, who asked her, "Is it possible that you were exposed to mold on your vacation?" YES! She'd identified visible mold in the hotel room where she was staying—in the bathroom and around the windows in the bedroom. She requested a new room, but the hotel was booked, so they did their best to clean the mold. That was a piece to the puzzle, which led Brielle's functional nutritionist to recommend she test her body for mold.

A urine test by Great Plains Laboratory (now operating as Mosaic Diagnostics) confirmed high levels of *Ochratoxin A* (OTA), *Aspergillus*, and *Penicillium*.

We've talked a bit about *Aspergillus* and *Penicillium*, but *Ochratoxin A* is a toxin produced by mold. While you can be exposed through contaminated food, the most common source of exposure comes from mold in buildings. It's a devastating toxin that studies suggest may be linked to kidney disease, decreased dopamine levels, neurological symptoms, oxidative damage to the brain, and neuro-degenerative diseases like Alzheimer's and Parkinson's.

The good news is that there are ways to cleanse the body and reduce the effects—like vitamins A, E, and C, as well as sauna treatments.

Brielle had actually been using an at-home sauna for two months prior to her mold test. After receiving the results, her functional nutritionist stated that her levels of mold would likely have been much higher had she not been using the sauna.

Her son was also tested for mold and had even higher levels than her tests showed. The doctor who read the results to Brielle said, "Your baby has been mold poisoned."

He had high levels of *Ochratoxin A*, *Aspergillus*, *Penicillium* (the same things Brielle tested high for) and *Stachybotrys chartarum*, with levels almost double the amount considered high. (The high range is greater than 12.61, and his level was 20.50.)

The doctor told Brielle the levels and strands of mold detected in her son have been shown to cause loss of speaking abilities in children, cognitive impairment, anxiety, and depression. Her son had been living in his room all nine years of his life. Months later he was also diagnosed with sensory processing disorder and ADHD, and he had a hard time attending school.

Her other son had significant food sensitivities identified at four weeks old. He would break out into body rashes, which lead to Brielle eliminating foods from her diet to determine the source of his sensitivities. She found dairy and legumes to be the culprit and cut these from her diet for two years while nursing.

Her youngest son outgrew those sensitivities, but at age five, he tested sensitive to eggs, and tests detected high levels of candida in his gut, which could lead to leaky gut if not managed properly.

The next question became: Where did this mold exposure come from?

She asked her mom about mold and, like me, found out a house she lived in for a year before college had mold that was eventually remediated, so perhaps the mold detected in her body was from past exposure. Brielle's functional nutritionist recommended starting the process of testing her home, with hopes of ruling that out, and suggested ImmunoLytics as an affordable first step to test the home for mold.

She tested her bedrooms, bathrooms, and living room/dining room/kitchen. High levels of mold were found in all but the living room/dining room/kitchen, including *Cladosporium* and *Penicillium*. The highest levels found were in the main bedroom and her son's room, followed by both bathrooms.

Brielle contacted a mold consultant to help locate the source of mold and explore what was required to remediate the mold. During this time, she moved a bigger air purifier into her son's room and kept his windows open, hoping that the mold exposure for the time being wouldn't have any further negative health impacts for her son.

But then he began to develop severe nosebleeds, where nickel- to quarter-sized blood clots were coming from his nose multiple times a day for four consecutive days. Brielle had him sleep in her other child's room (no mold detected on the ImmunoLytics mold test for this room), and the nosebleeds stopped. It was at this point she made the decision to move her family out of their home and into a rental home so that they could begin the renovation process.

I asked Brielle, just as I asked John (see Case Study #1 on page 27), what she would want to tell people about mold, given what she knows now. This is what she said:

> My experience has confirmed that mold is a health-compromising toxin that can have a variety of impacts to our long-term health. Our bodies are the only home we will have our entire life. It's worth it to determine whether mold is a root cause toxin affecting your health.

I also asked her to share what treatments she found most helpful in her process of home remediation, and she recommended the following:

- Home mold tests: Both she and I used the testing option from ImmunoLytics (https://immunolytics.com).

- Air filters: She's had the best luck with the Austin Air, AirDoctor, and Levoit brands.

- Portable home sauna: Hers came from Therasage (https://therasage.com).

- Electrolyte-replenishing water.

CALMING TOOLS FOR SUPPORT

You have come far in learning about mold and trying to wrap your head around the possibility of it being in your home.

The worry of having a moldy home is an intense experience! I know the overwhelm of uncovering mold and the questions that may be spinning around in your head:

How soon do I test my house?

How do I find the right doctor to help me test my body?

Is mold really what has been affecting my body and mind for so long?

What is the next best step?

All these questions are valid, and I have complete faith that you will be able to address them one day at a time with this book as a guide.

For now, let's come back to our growing toolbox so that we don't become too overwhelmed by the task at hand.

Let's pause with three calming tools for you to incorporate into your life.

TOOL #1: SELF-CARE ON STEROIDS

When I wrote this chapter, my son was sick. Very sick. We didn't know what he was sick with. We later found out that the grasses in Arizona were a major allergy trigger for him, which was exacerbated by the inflammation still present in his body as we continue to work on clearing the mold. We moved to Arizona in October 2021 and bought a home in April of 2022. As I mentioned, given my previous history, I tested our new home for mold before we purchased it. Nevertheless, despite my best efforts to create the safest and least toxic environment possible, something was going on that was causing him to have headaches, nausea, sinus congestion, and abdominal pain. In fact, I wrote the beginning of this chapter while watching him sleep in the emergency room at Phoenix Children's Hospital.

My world was full of doctors' appointments, emailing my son's teachers about missed days of school, acupuncture and chiropractor visits, specialist upon specialist, blood draws, X-rays, and of course, testing our home one more time for mold, just to make sure...

And therein lies the problem—not knowing how to fix his pain.

I was worried and exhausted. I remember how heavy my eyes felt as I typed the words on this page.

Self-care was a *must* for me in that moment, just as it is a must for you as you embark on this journey of mold discovery.

My self-care couldn't consist solely of massages and girls' trips, however. I had a responsibility to my child, and our family needed me to be present for the hardship we were facing.

At the same time, I needed a self-care plan if I was going to survive this trauma of having a sick child.

Believe it or not, writing was part of my self-care program. That is the cool thing about self-care—it can look different for all of us. Writing relaxes me, and I find it meditative. It is on my list of things that I do to care for myself. My list looks like this:

1. Eating a healthy meal
2. Writing
3. Gentle stretching
4. A foot soak
5. Lying down with my dog, my kiddos, or my husband
6. Meditation
7. Mindfulness practice
8. Reading inspirational books
9. Taking a quick nap
10. Calling a friend

In the space below, I encourage you to make a list of at least eight things that you can do for your own self-care practice.

_____

_____

_____

_____

_____

_____

_____

_____

Keep this list active, put it somewhere where you can see it, and set goals to achieve a few of the items from your list each week.

TOOL #2: JOURNAL WRITING PROMPTS

Sometimes the best medicine is to sit down and write, write, write. Write your feelings out until they aren't so strong in your mind. Let the worry scream onto the page instead of swirling around endlessly inside your head.

For this tool, you can grab a piece of blank paper and simply start writing or you can follow the journal prompts below:

Journal Prompt #1: What is your greatest fear about mold in your home?

Journal Prompt #2: Where in your body do you feel that fear most?

Journal Prompt #3: What story does your fear tell you?

Journal Prompt #4: If your fear was a small child or a favorite pet, what would you tell it?

Journal Prompt #5: Placing your hand on your heart, list three positive things that are happening today because of this moldy mess. Write them down.

Journal Prompt #6: What positive takeaway from this writing exercise would you like to remember for tomorrow? Write it down on a Post-It Note to hang in your house.

If it helps, you can set a timer for yourself, or you can just keep writing until you feel like you're "done" for the moment. Be gentle and try not to censor yourself—this is your chance to explore your feelings, and all of your feelings are valid.

TOOL #3: GRATITUDE FLOWER

People often fail to realize how chronic illness can impact your feelings of worthiness.

Having an active health condition can sometimes leave us feeling "less than," and it's easy to fall into the trap of blaming ourselves for not being able to just "push through" or "snap out of it." That mindset can chip away at our self-worth, which makes it even harder to invest in our well-being or even devote the time, money, and energy it takes to treat our bodies and our homes.

We need to see the good within ourselves so that we can see the importance and value in our own healing. Sometimes that means being proactive in reminding ourselves of our strengths and all the beauty within and around us.

In the center of the flower on the facing page, write your name. Then, in each of the petals, write something you appreciate about yourself or your life.

This active act of gratitude can help you cultivate self-esteem and self-worth. When we actively work on acknowledging and valuing these qualities, we quiet the voice of negativity that so many of us struggle with

CHAPTER 6

FINDING THE
MOLD INSIDE YOU

Knowing that you have mold in your home can be scary for sure. Finding out that you have mold in your body can be downright terrifying.

But it can also be freeing. Once you know that you have mold, you can get to work on treating yourself and making the necessary changes to having a healthier and happier YOU (and family).

When my daughter's functional medicine doctor recommended that we test her body for mold, I had to do a double take. At that point, I had no idea that you could test your body for mold exposure. I definitely didn't know that if you were exposed to mold, it could stay with you, even if you eliminated the source of the exposure.

The fact that my daughter, my son, my husband, and I all had results that suggested mold toxicity stopped me in my tracks. I knew it was highly unlikely that it was a coincidence, and those test results drove me to do the research and understand the options to have our home tested again; otherwise, we might still be living in that moldy home.

Testing your body for mold is important for a few reasons:

1. Your mold exposure may not be from your home. You may be getting mold from your office building, or your child might be getting mold from their school. Sadly, that's not as unlikely as you might think. In 2007, the EPA studied a sample of one hundred US office buildings and found

that 85 percent had water damage from a previous leak, and 45 percent had active leaks in the building.[30] In 1997, a survey of almost eight thousand US schools conducted by the General Accounting Office showed that 30 percent of those schools had plumbing issues and 27 percent had roof issues.[31]

Obviously those stats don't tell us how many of these buildings had mold growth, but they do show how many of the buildings that we spend time in every day could be at risk for mold. Testing your body can shed light on exposure you've received outside your current home. It is possible that you test your home for mold, and it comes back negative, but you find that you have mold in your body from a different exposure source. You want to know this so you don't miss getting the help you need.

2. Depending on the home test you use, you may not get an accurate reading—just like our family didn't the first time around. We found out the hard way that not every test is appropriate for every home or situation, and we learned that not every "expert" is created equal. Your home test might come back with no major red flags, like our first one did, even though you've got mold lurking and making you sick. Testing your body adds another layer of information and another data point. If your home test comes back negative for mold but your body testing comes back positive, you may need to keep digging. At the very least, it will give you the information you need to start a protocol to clear the mold, even if you are still working to pinpoint the source of your exposure.

3. If you do have mold in your home, you will still want to know how much mold is present in your body. This will help you in the future after you have cleaned the mold from your home (or moved) and done a detox protocol to rid your

body of mold. You will want to have a baseline mold reading so that you can retest to be sure you have properly cleared the mold from your body.

That baseline reading will also be important as you move forward. You'll be able to use that first reading to determine whether you've had another instance of mold exposure. If you notice an uptick in your symptoms and you have your body retested, you'll be able to see whether your numbers have continued to decline (in which case, you can feel more confident that you haven't been exposed to more mold) or have started to rise (in which case, you may want to have your home retested or start looking at other places where you might be exposed to mold outside your home).

The process of testing your body for mold can be a tricky one. You'll hopefully have a functional medicine doctor or naturopath who can help walk you through each step, but I think having an overview of the process helps us mentally prepare and helps us know what questions to ask the experts along the way.

WHAT YOU SHOULD KNOW ABOUT MOLD PRIOR TO TESTING

There are a few things you must know about mold before you begin testing.

Despite what many well-intentioned doctors may tell you, not all people with mold can detox it from the body by just moving out of the moldy environment. I have had this experience. At one appointment with one of my son's doctors, I pointed out that he'd had mold exposure, which I believed was contributing to his asthma.

She told me, "Oh, once you remove the source of the exposure, the mold clears out of your body on its own. It's like an allergy to cat dander; you just need to get it out of your environment."

I'm thankful that I had already done extensive research into mold so that I could stand firm in what I knew and advocate for my son. I told her that there were a number of studies that demonstrated this wasn't the case; that even once you cleared the mold from your environment, the mold in your body would stay with you.

She told me that those were the guidelines from her professional association, and that was what she followed.

Obviously, that was a frustrating experience. I'm hopeful that the medical associations will continue to review the growing number of studies about mold toxicity and update their guidelines.

You may find yourself in a similar conversation with a medical professional at some point in your mold journey, so here's what I want you to know: If you think/know you have mold in your environment, it is a good first step to remove yourself from the source of mold. However, you will still want to test your body for mold. If you test positive, begin a protocol to clear the mold (see Chapter Eight), and do another test in six months or a year so that you can compare your tests.

The reason mold does not simply leave the body the way a normal allergen, like cat dander, would, even when you eliminate the source, is that mold can dissolve in both fat and water.

Normally, toxins in the body would be filtered through the liver and other organs to help the body stay as healthy as possible. Mold, however, does not go through the classic detox process but rather gets reabsorbed back into the body before it gets a chance to exit.

In *Toxic*, Dr. Neil Nathan explains that "because mold toxins have the ability to dissolve into fats at one end and into water at the other, they essentially can move through any body tissue at will, as well as through all membranes ... insidiously forcing a wide variety of tissues to react to their toxic qualities."[32]

Nathan goes on to share that this does not happen to everyone because some people possess a gene to help with mold detoxification, however, "25 percent of people lack this gene and are predisposed to being affected by mold toxins; these are the patients we see with mold toxicity."[33]

In my experience with families who have mold exposure, every one has had to do some sort of treatment to help their bodies detoxify from mold. However, you can see in my family's story, and in the case studies in this book, that the degree to which one is affected by mold can vary from person to person.

Additionally, when you do get a mold test for yourself, you need to know that testing for mold in the body can be a little tricky, because sometimes you can have so much buildup of mold inside your body that it won't register on a test. There are, however, a few ways to test your body for mold and a few tricks for helping mold show up on those tests.

MOLD TESTS

There are a few different tests available for you. I have the most experience with the organic acids test, but I recently did a mycotoxin test for both my son and daughter to recheck the levels of mold in their bodies. So I'm sharing details on a couple of the labs that offer these test options—RealTime Laboratories and Mosaic Diagnostics.

I want to say up front that I don't have any affiliations with the companies I have suggested (meaning I don't get paid to refer you to them) for either the mycotoxin or organic acids test, and there are other companies that offer these tests. That said, these are two companies I either have direct personal experience with or have found doctors who I respect who use them.

Your strongest ally in this process will be a functional medicine doctor or a naturopath. Our family found a wonderful functional medicine doctor who has guided us through this complicated process, so much of my personal experience is working with this particular kind of medical professional.

A functional medicine doctor can help guide you in deciding which test will be best for your particular situation. They're also the experts in helping you interpret your results and deciding on a protocol to help you clear the mold from your body and start healing.

If you've never worked with a functional medicine doctor, the good news is that it's becoming more accessible for more people. If you don't have a functional medicine doctor in your local area, or if the functional medicine doctors in your area aren't affordable enough for your budget, you can work with professionals who live all across the country, thanks to the option of telehealth appointments.

The best place to start your search for a functional medicine doctor is the International Society for Environmentally Acquired Illnesses (ISEAI) database, which you can find at https://iseai.org/find-a-professional.

WORKING WITH A FUNCTIONAL MEDICINE DOCTOR

According to Functional Medicine SF, which was founded by Stephanie Daniel, DO (an amazing functional medicine doctor),

- "Functional medicine involves understanding the origins, prevention, and treatment of complex, chronic disease."
- Functional medicine is an integrative, science-based healthcare approach.
- Functional medicine is patient-centered care.[34]

While we opted to work with a functional medicine doctor, I know many families who have found incredible support and guidance from naturopathic doctors. Naturopaths, like functional medicine doctors, take a holistic, patient-centered approach to healthcare. Naturopaths focus on using evidence-based natural interventions, specifically leveraging the body's own natural ability to heal itself, to promote overall health and address illness.[35]

And there's one more caveat I have to share here. Unfortunately, there is no "perfect" test that I am aware of when it comes to testing for mold in your body. And it is important to note that not all mold will show up on a test.

Nathan also shares that many mold-toxic patients have difficulty showing a positive mold result on their test, even though mold is present in their body. The reason for this, says Nathan, is that "the toxins have literally poisoned the very systems needed to remove these toxins ... making it difficult for patients to mobilize those toxins (even though they are there) sufficiently to make the test positive."[36]

Nathan's solution? To use oral glutathione, 500 milligrams twice daily, for a week and then collecting a urine sample.

You can also use a sauna or hot tub to help your body sweat more prior to a test because that will help your body detox and thus show more of your toxins on the test. And if you're working with a functional medicine doctor, they may be able to build a protocol that will make the mold easier to detect in your system.

A word of caution here, though, as pushing yourself too hard when you have mold toxicity can really aggravate your symptoms.

This is a tricky dance, and if you can find a professional to guide you in your tests and interpret your results, that will be the best possible scenario for you!

Below are some recommended tests for you to pick from and discuss with your doctor.

MYCOTOXIN TESTING

Mycotoxins are toxic chemicals found in mold. Mycotoxin testing can be a blood test or a urine test, depending on which company you use.

RealTime Labs

RealTime Labs' mycotoxin test detects sixteen different mycotoxins, including nine macrocyclic trichothecenes, which are secondary metabolites produced by fungi that can cause damage to our bodies at the cellular and subcellular level.[37]

This test can be ordered in twenty-six states without needing the authorization of a doctor, and then you can schedule a consultation with a member of their team if you would like to go over the results.

You can learn more on their website: https://realtimelab.com/mycotoxin-testing.

Mosaic Diagnostics

Personally, I used Mosaic Diagnostics (which, at the time, was operating as Great Plains Laboratory) for my family's mold testing. They offer a "MycoTOX" test that screens for eleven different mycotoxins, from forty species of mold, in one urine sample.

You are not able to order a test kit directly through Mosaic Diagnostics (you must have a physician order one on your behalf). You may, however, contact their customer service team, and they will refer you to a physician in your area if you need help finding one.

You can learn more on their website: https://mosaicdx.com/test/mycotox-profile.

These are just two of the mycotoxin testing resources available to you. Again, if you have a healthcare practitioner, you can see what testing service they recommend for you.

ORGANIC ACIDS TEST (OAT)

Organic acids testing is used by many medical professionals to identify nutrient deficiencies but is also very effective in flagging mold toxicity because mold mycotoxins also produce organic acids:

> The Organic Acids Test (OAT) offers a comprehensive metabolic snapshot of a patient's overall health with 76

markers. It provides an accurate evaluation of intestinal yeast and bacteria. Abnormally high levels of these microorganisms can cause or worsen behavior disorders, hyperactivity, movement disorders, fatigue, and immune function. Many people with chronic illnesses and neurological disorders often excrete several abnormal organic acids in their urine.[38]

This was the test that our family used when we were first diagnosed for mold, and I have seen others achieve successful results with this test as well.

You can find more providers online that offer the OAT, or you can connect with a physician and see where they suggest you get a test done.

SLOW AND STEADY

Whatever test you choose, be slow and steady when you get the results back.

Many people with mold toxicity cannot quickly clear mold out of their bodies. The longer the mold has been with you and the sicker you have become, the more time it may take to completely heal.

My personal story around clearing mold from my body has been very slow. It was through the organic acids test that I discovered I had mold (twenty years after my first exposure in high school!).

My report stated that I had:

- High yeast/fungal metabolites
- High DHPPA (3,4 dihydroxyphenylpropionic acid)
- Homovanillic acid (HVA) levels (33) below the mean
- Vanillylmandelic acid (VMA) levels (34) below the mean
- High HVA/VMA ratio (35)

- 5-hydroxyindoleacetic acid (5-HIAA) (38) levels below the mean
- Pyridoxic acid (vitamin B6) levels below the mean (51)
- Pantothenic acid (vitamin B5) levels below the mean (52)
- Ascorbic acid (vitamin C) levels below the mean (54)

I wanted to just take a binder for mold and clear it out. I wanted to simply take more vitamin B and fix everything right away, but I quickly found out that my body was so sensitive from prolonged mold exposure that I had to slowly work my way up to taking the best medicines for me. (I am still working on this even as I write this book!)

I also learned that I had MCAS, which compounded my challenges in trying to detox my body. MCAS is a topic that I have been alluding to throughout this book, and we're going to dive into it in the next chapter.

ONGOING TESTING

Chances are good that testing is not going to be a one-time deal if you do discover mold in your body through your first test.

You'll work with your functional medicine doctor to decide when retests are a good idea, but I don't want you to worry that you're going to be opening yourself up to constant testing.

For the most part, during the healing phase, which we'll talk about in the next chapter, you'll be tracking your symptoms and how you feel physically and mentally.

In my case, I've been tested three times in total thus far, but I know that I'm making progress because my symptoms have decreased significantly and I am feeling so much better. I have more energy. My brain fog is clearing. My anxiety is low. I feel happier.

We've followed the same track with my kids. They've had a few bumps in the road along the way, and we retested just to check their progress. When we moved into our new home, testing their bodies was another way to confirm the home test we did, which came back negative for mold. If their levels of mold had increased after being on a healing protocol, I would have known there was a new exposure. I am happy to report that their levels of mold are much better, and they are making great gains!

Clare's case study gives you another example of someone whose mold testing on their body was the first indication that they'd been exposed to mold, which prompted them to test their home, and it shows how easily mold poisoning can be mistaken for another ailment.

Case Study #5: Internal Tremors and a Misdiagnosis of MS

Clare's physical symptoms began in 2013 when she first started noticing strange internal tremors or vibrations, as if her body were vibrating or shaking even though she wasn't physically moving. Over time, she also noticed additional symptoms of shooting joint pain, tingling of the hands and feet, and serious brain fog as well as other neurological tics.

Eventually, in October of 2021, she had an episode of internal tremors that felt like she was in an earthquake, although nothing was moving around her. This finally sparked her to figure out what was going on.

Her doctor told her these symptoms were most likely the early stages of multiple sclerosis (MS). However, after a normal brain MRI, a neurologist specializing in MS told her that she had something strange happening, but it wasn't the disease.

Frustrated with the lack of answers, she visited a naturopathic doctor with a background in regenerative medicine. Within fifteen minutes of meeting with her, the doctor told Clare her symptoms sounded like mold toxicity.

Through her doctor, she completed a MycoTOX profile to determine whether she had mold in her body. The results were astounding. She was off the charts in levels of citrinin (dihydrocitrinone DHC) from multiple mold species including *Aspergillus*, *Penicillium*, and *Monascus*, at 472.17 ng/g creatinine—with the normal range being <25 ng/g creatinine. She also showed high levels of *Ochratoxin A* from *Aspergillus*. A normal range is <7.5 ng/g creatinine; hers was 15.36 ng/g creatinine. I realize that was a lot of information (you can see why it's so important to work with a functional medicine doctor or a naturopath, who can help you interpret your results), but here's the big takeaway: Clare's tests showed an overwhelming amount of mold in her system.

She had her home tested by a company using an Air-O-Cell sample test, which came back showing dangerous levels of *Aspergillus* and *Penicillium*. In other words, she had toxic mold growing in her house.

In April of 2022, Clare left her home so a professional mold neutralization and remediation company could neutralize the mold, clean it out, and rebuild the areas that had been affected.

I asked Clare what she would want to tell people about mold given what she knows now. Clare said:

> I wish I had put all the pieces of the puzzle together sooner. It took almost ten years before I got to the bottom of it. I minimized my condition and explained away my symptoms until finally, I couldn't make excuses for them anymore. I realized how I was feeling wasn't normal and I needed to find out why. It just never occurred to me that I could have mold in my body.

Looking back, my husband has had neurological symptoms too and all three of my kids have had respiratory issues. I'm so glad we found the issue and fixed it before any more harm was done.

Clare is still in the beginning stages of detoxifying the mold from her body and hasn't started any treatments yet. She is grateful to have the mold out of her home though (a big first step)!

Clare is an amazing example of someone who was in tune with her body and the health of her family. She knew that something was off but couldn't quite put her finger on it. After connecting with a doctor who understood mold, she began to uncover the hidden problems that were greatly impacting her so she could set out on a path to regain her health.

CALMING TOOLS FOR SUPPORT

Finding a calm state in the chaos of mold is crucial for the healing of your body and mind. These tools are just as important as the tests and the detox medicine/vitamins you will take (which we'll discuss more in the next Chapter Eight).

Chances are good that whether you start by testing your home or your body for mold, if you've decided to make the investment in testing, you are probably suffering from some health challenges that you want to resolve. They are likely negatively impacting your day-to-day life, and you may have already spent a lot of time, energy, and money on trying to find answers and solutions.

All of that can sometimes make it difficult to treat your body with the compassion it deserves and needs.

These tools are designed to help you reconnect with and heal your body, which needs nurturing and patience as you move through the process of detoxifying.

I hope you embrace these tools and incorporate them into your daily routine!

TOOL #1: BRING IT BACK TO THE BREATH

What if you could move from your sympathetic nervous system (the one in charge of fear and stress, which gets activated when you perceive a threat) into your parasympathetic nervous system (the part of your body that helps you root into calm)?

What if you could do that by just using your body and do it all for FREE?

"Take a deep breath" isn't just an annoying saying that well-intentioned people say to you when you are upset and overwhelmed.

Scientists have studied the breath, and they now know that by taking at least six deep breaths, you can switch from your sympathetic nervous system to your parasympathetic nervous system.

Begin by straightening your back and placing your hand on your stomach or heart.

Take in a deep inhale through your nose for a count of *1, 2, 3.*

Follow that with a soft exhale through your mouth for a count of *1, 2, 3.*

Keep repeating this until you feel more regulated and can move into one of the other tools from this book.

TOOL #2: SELF-PORTRAIT

This is an activity that I have done in my teaching with children, and I have since begun to use it for adults. It is another tool that gets your mind out of what is wrong and into what is going well.

On a piece of paper, draw an outline of your head. This can be as simplistic as you would like to make it. My portrait is the outline of my head, ears, hair, and neck.

Don't add any facial features. Keep that space open because you are going to write all the things you appreciate about yourself inside your head.

My list includes that I am empathetic; I don't give up easily; I am resilient; I am a good friend; I am a good mother, wife, and daughter; I am kind; I try my best; and the list goes on.

There is a lot of good that you are doing in your life, and it is time to put the magnifying glass over what is going well.

TOOL #3: NEW NEURAL PATHWAYS

Neuroscience research has found that we can create new neural pathways to change our brains with intentional activities. This rewiring can even change our genetic set point.

When I work with children, I call this "making a new memory."

I explain that we can change how we feel about an activity, a friend, a type of food, etc., by the way we think about and experience things that have been unfavorable in the past. When children (and adults!) "make a new memory," they do this by turning the negative into the positive.

Answer the questions below to begin on your path of making new neural pathways in your brain about your health.

Question #1: When your mind tells you that you will never heal and you are stuck in this situation, what kind statement can you tell yourself?

_____

_____

_____

Question #2: What are three things that you love about yourself just as you are, now?

_____

_____

_____

Question #3: What can you do today to support your physical and mental health?

_____

_____

_____

PART 3

HEALING FROM MOLD

CHAPTER 7
Histamines and MCAS

I waded through a river of water that was separating one side of the beach from the other. Sea turtles were surrounding our family as they made their way up the shore of their Costa Rican breeding ground.

My two children couldn't believe all the eggs that the turtles were laying. The sun began to set, and the stars came out so vibrantly that they appeared to be peering at the turtles right alongside us.

It was the summer of 2019—before we'd discovered the mold lurking in our home—and our second trip to Costa Rica. I knew this moment was magical. I willed myself to calm down. I pleaded that my mind would stop racing with dread. I couldn't seem to stop the worried mind that night. In fact, I seemed to be getting more anxious by the day as we vacationed in one of my favorite places on Earth.

I couldn't understand why my anxiety had gotten so bad since the beginning of the trip. Was I just out of my element? No, probably not. I am an avid traveler, and I love a good adventure.

Was I about to get my period? No, not that either.

And then it dawned on me!

Every day since our arrival, I had been taking the kids to the local market and getting a green drink made with kale, coconut water, spinach, and tropical fruit. Back home, I had always had a morning smoothie for breakfast, and since I didn't have access to my usual ingredients, I had been substituting with this drink.

I decided to take a break from the juice. After all, I couldn't feel any worse than I did at that moment.

Within two days, I was feeling better mentally. What I didn't understand then, but understand now, is that everything in that juice was high in histamines, which was exacerbating the inflammation mold had caused in my body.

IS YOUR DIET MAKING YOUR MOLD TOXICITY WORSE?

Being conscious about the foods I was eating has been a lifelong practice for me, but it became a major focal point in my life after my first mold exposure in high school.

I told you that my food sensitivities started long before my first mold exposure, but when my mom and I moved into that apartment when I was in eighth grade, my stomach issues and sensitivities skyrocketed to the point that they were absolutely debilitating. When I left for college, I had gotten some relief by moving out of that home.

Still, though, my stomach pain would arrive at the most inconvenient times, and my mood was lower than it had been in years.

When I was in college, I was seeing an acupuncturist (which I found helpful) and would sometimes visit the doctor at student health. But really, I was in a limbo. My health wasn't great, but I had agreed to all the tests my traditional doctor could think of. We knew I didn't have Crohn's disease, but no one could tell me what was making me so sick. By that point, it had begun to feel like I had no options other than to accept a life of poor health.

In an unexpected miracle, my mom came to visit me in my yellow rental home and brought me a book about wheat and gluten allergies. I will never forget the saffron-colored cover and the lists of gluten-containing ingredients listed within the pages of the book.

I read through that book more times than I have read through any other book in my life and related to almost everything written about gluten and wheat allergies.

I started to learn more about gluten intolerance and discovered that it can cause:

- bloating in the abdomen,
- anxiety,
- anemia,
- fatigue,
- diarrhea,
- constipation,
- nausea,
- headaches,
- and more.

That was around 2005, and almost no one knew what gluten was! Really, many people still don't know about gluten and the gut-brain connection.

Today there's a lot of research being done into gluten, and I can't lay all of it out for you here in this book. But if you think you have mold, it's very important to understand how diet affects the body. And you will really want to consider how much gluten you are consuming, as gluten can be inflammatory and make your symptoms of mold toxicity worse.

According to Dr. Susan Tanner, who specializes in environmental medicine and chronic illness, and Eric Hosford, a registered nurse:

> The specifics of gluten's role in yeast overgrowth are much more than its prevalence as an allergen or in the fact that it is found in foods that Candida albicans (yeast) consumes. Gluten itself is difficult for the body to break down and, simultaneously, has a relatively large molecular structure. Supposing GI damage has already occurred via yeast overgrowth (sometimes because of the presence

of mycotoxins from mold in the GI tract—either from environmental mold exposure or from consuming mycotoxin-laden foods) in the intestines, absorption of these oversized and undigested glutens can be a process that is further destructive to the intestinal lining.[39]

So we're not going to dive into the detailed ins and outs of gluten. Instead, we're going to focus on how gluten's effect on the body can exacerbate the consequences of mold your body is experiencing and make it harder to clear out the mold and heal. But to do that, I need to give you a crash course in gluten.

WHAT EXACTLY IS GLUTEN?

If you were like me back in 2005 when my mom brought me that book that would change my life, you may not know what gluten is either.

In short, gluten is a term for the proteins that are found in plants. Some of these plants and their derivatives include:

- Wheat
- Wheat derivatives (wheat berries, durum, semolina, farina, farro, graham flour, etc.)
- Spelt
- Rye
- Barley
- Malt
- Brewer's yeast

There can also be wheat products within many foods that we eat. For example, most soy sauce has wheat in it.

I read the book about how to be gluten free way before things were labeled at your local grocery store as "gluten-free." In fact, I remember going out to dinner at an Italian restaurant when I had first decided to stop eating gluten.

It was one of my favorite places to eat—overlooking the small creek that ran through our town, sparkly lights, and the laughter and loudness of college students. I looked at the menu, perplexed about what to order. I knew that pizza was a no go, but could I eat pasta? What about gnocchi?

I told the waiter that I was allergic to gluten and asked him if he knew if there was gluten in the pasta. He had no idea and went back to ask the head chef. For those of you who don't know, pasta *absolutely* has gluten (unless you buy it gluten free). The head chef didn't know that and confidently told me that the pasta only had flour and water in it (again, flour = gluten).

I ate the pasta and hours later had a horrible stomachache.

That was how the bumpy path of learning to be gluten free began for me.

After that night at the Italian restaurant, I went to the local copy store and made multiple lists of all the ingredients that contained gluten. I carried that list around in my purse for years and would take it to the grocery store and read through each ingredient on all the packaged food.

What I found, though, was that as I became more and more gluten free, my intestinal issues began to decrease! Woot woot!

DECREASING GLUTEN (AND OTHER INFLAMMATORY FOODS) IN YOUR LIFE

You are going to want to examine the role of gluten and other inflammatory foods in your life if you believe that you have mold toxicity.

For me, my sensitivity to gluten was another clue to my mold illness, which I didn't understand then but can see clearly now.

The gut is more powerful than you think. Scientists now know that within the walls of your digestive tract is your "second brain," so-called because so much of our mood and mental health can be traced back to its microbiota. According to Johns Hopkins Medicine, "scientists call this little brain the *enteric nervous system* (ENS). And it's not so little. The ENS is two thin layers of more than 100 million nerve cells lining your gastrointestinal tract from esophagus to rectum."[40]

Many doctors used to think that if a person was anxious or depressed, those emotions could trigger digestive problems like irritable bowel syndrome and constipation. That is true! However, what scientists have now learned is that what we eat can also trigger how we feel!

If you have mold toxicity, then you most likely are feeling a host of physical and psychological symptoms. And what you are eating can further exacerbate the problem (especially if what you are eating is inflammatory).

Inflammatory foods include gluten, sugar, processed foods and meats, fried foods, and the list goes on! We call these foods "inflammatory" because they cause inflammation in the body. Mold can cause a chronic inflammatory response in the body, and thus we want to be careful of adding extra inflammation.

Be mindful about the foods you eat and how they may or may not affect the way you feel.

Grab a journal, if you would like to, and record what you eat for a week or two and how you feel after each meal. Notice any correlations?

LET'S TALK ABOUT HISTAMINES

When my mom brought me that book about gluten-free eating, I got some major relief from my pain. I was able to begin to expand my diet a bit and gained back some healthy weight, and even with a few setbacks, I continued to feel better the longer I stuck to my gluten-free diet.

When my husband and I moved into our home in Northern California in 2009, though, I started to experience enough new physical and mental setbacks that I thought a change in my diet might be due. I knew gluten was still out for me, but there were lots of protocols that allowed me to abstain from gluten while focusing on other aspects of my diet.

I tried every diet I could think of—vegan, vegetarian, Paleo, keto, the Medical Medium diet of low protein and lots of fruit ... but nothing ever seemed to fully solve my physical symptoms.

In fact, many of the diets that were supposed to be helpful for maintaining a healthy gut had the opposite effect on me.

For example, it is common to take vitamin B for fatigue, but whenever I took vitamin B, I would become more fatigued and emotional!

Research shows the benefits of fermented foods on the gut. Yet, when I would try to eat fermented foods, I would feel lethargic and out of sorts physically and mentally. The same would be true for eating foods such as bone broths, mushrooms, strawberries, and stews.

It turns out that the histamines in these foods were behind the new troubles I experienced when I was reexposed to mold in our home.

Histamine is what your body produces to protect you from your typical allergens like pollen, pet hair, and dust; histamines also help get rid of these allergens, hence why we get symptoms such as itchy and watery eyes, coughing and sneezing, and all that fun stuff.

If you're allergic to mold and have mold toxicity, well, histamines found in food most likely aren't going to be your friend. Your body is likely already flooded with histamines, which are being released to combat the mold in your system.

To put it very simply, you already have enough histamines in your system; you don't want to introduce more from outside sources. Your symptoms could be magnified, especially if you're consuming histamine-rich foods.

From my observations after hearing other people's experiences, we can self-diagnose or be misdiagnosed by a doctor or health-related expert and end up on diets in which we're consuming foods high in histamine, which only exacerbates our mold toxicity condition (if we have it). Think Paleo diet, keto diet, Whole30—all of these contain ingredients on the "yes" list that are high in histamine.

I really want to stress that many of these high-histamine foods are nutritious and healthy, which can make the process even harder to navigate. Lots of us know that we should be careful about too much refined sugar, salt, and trans fat; almost every nutritionist, doctor, and diet guru will agree, and that guidance is everywhere.

But things like bananas, bone-broth soup, and walnuts? Those kinds of ingredients are go-tos in most diets! Who could imagine you'd want to avoid them? In fact, many experts would tell you to load up on nutrient-dense foods like these.

Here's the catch: If you have mold toxicity, and food sensitivities and intestinal/digestive issues are among your symptoms, histamine-heavy foods will probably be too taxing for your body. Elevated histamine levels are just one of many symptoms mold toxicity can produce, though these symptoms can differ greatly from person to person.

As you might remember from Part One of this book, one of my "stops" on the long carousel ride of doctors and specialists I visited was a doctor who told seventeen-year-old me that I had Crohn's disease (due to my intestinal problems)—one of many misdiagnoses to come. With that misdiagnosis, I was instructed to go on an all-white diet (remember?!)—white bread, white chicken breast, white rice, white potatoes, white cream of wheat, you name it.

And the kicker was that these foods were *all* high in histamine (and gluten too ... talk about a double whammy!). This new diet was supposed to help me better digest food and reduce the symptoms of my so-called Crohn's disease, right?

Of course, not knowing I had mold toxicity at the time, it did the exact opposite: My body was flooded with foods that were hard for me to tolerate.

Because of the minimal support and expertise in the realm of mold toxicity, it was difficult for me to find the treatment I needed. Histamine-loaded foods did not help my mold toxicity, which was prolonged for years, and that's why I want to inform you on this topic, so that you can avoid the mistakes I (and many others) have made while trying to fix our health problems. This brings me to another topic I would like to discuss: changes in your diet and lifestyle.

TIPS FOR DECREASING YOUR HISTAMINE INTAKE

Start to notice whether you experience physical or physiological symptoms when you eat food high in histamine. If we think of histamines as a bucket, then you may notice that you can tolerate one cup of yogurt just fine. Yet when you eat a few servings, you don't feel so great. It's important to take notice of these patterns. Have a food journal on hand to help you keep track of these times. See if you are overflowing your histamine bucket!

As you now know, an all-white diet is less than ideal (for both general health and specifically for mold toxicity), but did you know that smoked meats and dried fruits are also high in histamine? Unfortunately, foods and diets that can be very good and healing for the gut for most people may have the opposite effect on you if you have mold toxicity.

I have noticed a trend of people hopping onto the Paleo or keto diets, with the hope that their symptoms would resolve. Although many people may experience positive and thriving results, those diets actually can make mold toxicity conditions *worse* because of the histamine-heavy foods. I'm not saying you can't be on those diets, because you *can* get crafty and replace certain suggested foods with others that are low in histamine. Whatever diet you choose or foods you buy, it's important to focus on eating healthy, fulfilling meals that will help heal you from mold toxicity symptoms.

Please note that choosing a diet that avoids histamines for a prolonged period is *not* beneficial or healthy. What you'll notice is that a lot of the foods on the following list are healthy and nutrient dense, which you want in your diet. So it's best not to follow a low-histamine diet longer than you must. Use this list temporarily as a gauge, and see how you feel while you are working to clear the mold out of your body. My experience is that the more mold I have cleared from my system, and the more my body has stabilized, the better able I am to tolerate foods that are high in histamines.

Now let me share the list of foods to avoid and foods to eat to help make the healing journey easier to "digest," so to speak!

FOODS HIGH IN HISTAMINES OR THAT TRIGGER HISTAMINES

- Processed foods
- Dates
- Grapes
- Citrus fruits
- Papaya
- Plums
- Bananas
- Prunes
- Raisins
- Dried fruits
- Coconut (depending on how your body reacts)
- Pickled and preserved vegetables (such as olives, onions, pickles, sauerkraut, truffles, and kimchi)
- Sea vegetables and algae (such as spirulina, seaweed, nori, dulce, kelp, and wakame)
- Shellfish
- Fish that isn't fresh (such as anchovies, sardines, herring, and mackerel)
- Preserved meat (such as bacon, caviar, ham, salami, smoked meat, and vacuumed-packed meat)
- Baker's yeast
- Wheat germ
- Peanuts
- Walnuts (other nuts can be triggering too, so see how your body reacts.)

- Beans (depending on how your body reacts)
- Cow's milk (can include food products made with cow's milk, such as yogurt)
- Cow's cheese
- Aged cheese
- Alcohol (beer, champagne, cider, wine, liquors)
- Energy drinks
- Chocolate
- Sugar
- Soy milk
- Soda
- Gluten
- Vinegar
- The "all-white diet"[41]

YOUR GO-TO LIST OF FOODS TO EAT

What we eat can truly help us with our health and make our symptoms less intense. My advice to you is to approach a low-histamine diet by eating unprocessed and high-quality foods. Here's a list of some examples of low-histamine foods you can include in your everyday snacks and meals:

- Fresh poultry (can be cooled or frozen)
- Fresh beef (see if you can tolerate this)
- Fresh turkey
- Fresh veal
- Fresh bison
- Halibut, salmon, cod (fresh or frozen)
- Fresh fruits and fruit juices that are low in histamines—e.g., apples, apricots, blackberries, blueberries, cantaloupe, cherries, pears, nectarines, peaches, pomegranates[42]
- Fresh vegetables
- Some grains (think rice, quinoa, rice noodles, amaranth, millet, potato starch)
- Fresh pasteurized milk and milk products (see how you tolerate this)

- Milk substitutes like goat's milk and sheep's milk (see how you tolerate this)
- Butter and ghee
- Most cooking oils
- Most leafy herbs
- Herbal teas (except black and mate teas)
- Raw honey and maple syrup
- Seeds (chia, flax, hemp, poppy, sesame)

This list is a good jumping-off point, but please keep looking into low-histamine food options if you find that this diet change is helping you in your healing process.

If you do notice that you are making progress with a low-histamine diet, you might also want to dig in more and learn about MCAS.

I've mentioned MCAS throughout the book as one of the many consequences of mold toxicity I experienced, so let's finally take a deeper look at what it is and how it's connected to mold.

MY HEALING PROTOCOL

When I first began working with my functional medicine doctor, most foods felt unsafe. I had just tried the Medical Medium diet of celery juice, lots of fruit, small amounts of meat, and daily detoxing in the sauna.

At first, I had an initial rush of feeling great. I thought I had found the diet that was going to fix my ailments (this time I was sure of it), only to be let down again when my symptoms of brain fog, anxiety, fatigue, and low mood came back to hit me like a runaway train.

Disappointed and frustrated, I told my new functional medicine doctor about the foods I was eating and my reaction to them.

"That is because you have mast cell activation syndrome along with lots of mold," my functional medicine doctor told me.

Mast cell what? I thought. In all my years of research, I had never heard of it.

Honestly though, learning about MCAS in addition to my mold toxicity quickly became a relief. It was nice to finally understand the root cause of my poor health!

WHAT IS MAST CELL ACTIVATION SYNDROME?

When our bodies are exposed to toxins (or substances our bodies believe are toxins), they send out mast cells, which are a type of white blood cell and part of our natural immune system, to combat the foreign substance and keep it from making us sick. They do that by releasing chemicals, like histamine, which creates inflammation to protect that area of your body.

In spite of the fact that this inflammation can make us uncomfortable—think of something like the swelling that develops around a bee sting or the stuffy nose you get during a cold—those mast cells are actually doing their job to keep the foreign agents from settling into our bodies and making us even sicker.

For people who develop MCAS, though, those mast cells can actually start to cause physical damage.

Beth O'Hara summarizes MCAS perfectly on her website. She says:

> Mast Cell Activation Syndrome is a condition where important immune cells called mast cells have gotten out of control.
>
> When you have MCAS, your mast cells get overly sensitive. They become overly responsive.
>
> With overly responsive mast cells, you may be immediately sensitive to things that don't seem to affect others.

You may also respond to things that should be safe (like food)![43]

O'Hara points out that MCAS can be triggered by several factors, one of them being *toxic mold exposure*.[44]

Learning about my MCAS was another important piece of the puzzle in understanding why my body had such clear negative reactions to high-histamine foods. My mast cells had gone rogue and were releasing higher than normal levels of histamine when they were triggered.

The beginning stages of clearing my body of mold came through stabilizing myself with a low-histamine diet. It was clear to me that there was a whole list of foods that made my physical and psychological symptoms worse, but I never understood why until I learned about MCAS and overdoing it with high-histamine foods.

In the space below, make notes about foods you may have eaten recently that are on the histamine list above and how you felt after eating them. Do you think you might be affected by histamines, too?

_____

_____

_____

_____

_____

_____

If you want to learn more about MCAS, I encourage you to check out Beth O'Hara's website: https://mastcell360.com.

HONORING THE GUT-BRAIN CONNECTION

One of the most challenging parts of the healing process is navigating the ups and downs and setbacks without catastrophizing or telling yourself that all your progress has been lost.

In 2022, as I was writing this section, two years after we first started the process of detoxing our family's bodies (more details on the protocols for this process in Chapter Eight), my daughter came home with a virus after a family vacation to Colorado.

In the wake of that experience of being sick, her fear and worry suddenly came rushing back.

Of course, this raised all kinds of red flags in my mind. Had she been exposed to mold from a new source? Had her healing taken a backstep?

I immediately reached out to our functional medicine doctor. She told me that even though the protocol we were using had cleared a lot of mold from my daughter's body, she still has a sensitive microbiome in her gut. We know that there's a strong brain-gut connection, and the microbial consequences of getting a virus dysregulated her system again.

The disruption to her microbiome caused her nervousness to flare.

This time, though, we found that the hydroxyzine that she'd taken in the past didn't help calm her down like it had before. We weren't dealing with the same kind of inflammation from mold as when we'd first discovered it in her body.

Just to be sure though, we decided to order a new mycotoxin test. While she still had elevated markers for mold, they were trending downward as we slowly cleared mold from her body. That set my mama mind at ease, knowing that she hadn't been exposed again, and we weren't starting from square one.

This time she needed a more therapeutic approach to help her emotionally regulate. We used mindfulness tools and a session of EMDR therapy to help her—things that hadn't helped nearly as much when she'd been at the height of her mold toxicity because the mold had made it too hard for her nervous system to calm down—and she began to stabilize once more.

It's the middle of summer 2023 today, as I'm putting the finishing touches on this book, and I can now happily report that we have had more viruses in our home since that one on the heels of our Colorado trip, and none of them have triggered a spike of emotional dysregulation for my daughter, and her last mold test was almost completely clear!

I tell you this story so that you're not surprised if you encounter moments during your healing journey that feel scary, ones that make you worry and wonder.

Even when they look and feel like those symptoms you've been gradually leaving behind, know that you are still moving forward and taking the steps you need to heal your body. When in doubt, touch base with the professionals you're working with and see if having a follow-up test is advisable, even if it's only for your own peace of mind.

I also tell you this story to encourage you to hold on to your HOPE. Seeing my daughter's progress is a constant reminder for me that my body will heal in time as well (and so will yours)!

CALMING TOOLS FOR SUPPORT

I hope that you are beginning to feel a bit more empowered, more hopeful about the progress you can make and less alone. Here are three more tools to incorporate into your healing process. I am so proud of all your hard work.

Remember, YOU CAN DO HARD THINGS,
and
YOU WILL GET BETTER!

TOOL #1: GO ON A WALK

Sometimes we just need to get out of our head by getting outside in nature.

One of my favorite quotes comes from the nineteenth-century environmentalist John Muir: "I only went out for a walk and finally concluded to stay out until sundown, for going out, I found, was really going in."

Taking a walk (even if you can only go for a short period of time) is a great way to change your thought patterns and support your emotional needs.

TOOL #2: MINDFUL EATING

Mindful eating is an easy and simple way to bring the power of observation to something we do throughout the day. It is also very helpful for those of us who have food sensitivities because it relaxes the digestive tract as we eat our food.

According to Jan Chozen Bays, author of *Mindful Eating*, "mindful eating involves paying full attention to the experience of eating and drinking, both inside and outside the body."[45]

While we eat, we pay attention to our food before we put it in our mouths. What does it look like? What does it smell like? How does it feel if we are touching it with our hands?

Then, as we take a bite, we notice the way it feels as we chew each piece of food.

For me, I am often surprised how much sweeter a piece of fruit will taste when I bring mindful awareness to it.

How do you feel as you eat? Notice whether your stomach is getting full, or whether you need to take a break.

If you notice that your mind is getting distracted during this practice, don't judge yourself, but rather return your attention again to the taste, smells, textures, and feelings of your food.

TOOL #3: DRAW IT OUT

Expressing ourselves through art can allow us to communicate things that are difficult to put into words. It also gives us an opportunity to reflect on and identify what we are feeling—taking intangible internal emotions and making them external and visible so that we can create some space from them.

In the space on the next page, draw your feelings. Get out the pencils, crayons, markers, and any other tools you have lying around, and let your emotions hit the page.

DRAW IT OUT POSTER

CHAPTER 8

DETOXING FROM MOLD TOXICITY

As I write this chapter, it is a late hour for me (10:01 p.m., to be exact) on a warm summer night here in Arizona (91 degrees Fahrenheit). I am taken aback by the pain-free life I have today.

I just turned thirty-nine years old seventeen days ago, and yet it seems like yesterday that I was living in that apartment with my mom, running through the streets of San Luis Obispo, and then moving into my first home with my husband. It doesn't seem that long ago that there was all that good coupled with countless days of excruciating pain.

Yet here I am typing this chapter about detoxing mold from my body. Offering a voice to the story of mold, a story that is too often untold. Hoping that these words will inspire me to continue my path of healing for myself and my family, while also serving as a guide in your healing process. Feeling hopeful about my health (and yours).

As I turn off the air-conditioning and let the stillness of night flood the room, I think of all the pieces of the puzzle that have come together to help me heal. I feel better today than I have in years. It is with a heart full of gratitude that I share with you how I have healed thus far.

TREATING MCAS AND MOLD TOXICITY

My first order of business in treating my MCAS and mold toxicity was to make sure that I had frequent and healthy bowel movements.

Having healthy bowel habits means that you are having a consistent bowel movement at least once a day, fully emptying your bowels when you do, not having to strain or wait for long periods of time on the toilet, and that your stool is soft and easily passed (without crossing the line into watery diarrhea).

In order to have healthy daily bowel movements, I have found that taking magnesium each night before bed is extremely helpful. I like the Pure Encapsulations Magnesium Citrate.

Once my bowel movements were consistently healthy, I worked on stabilizing my mast cells by eating a low-histamine diet and beginning an antihistamine.

Now, let's be clear here. I do not believe that all antihistamines are created equal. Nor was it my experience with my daughter that she needed to be on an antihistamine for a long period of time.

In fact, some research suggests that taking a daily antihistamine can actually make MCAS worse because the antihistamine can cause your body to produce more histamines in the long run (confusing, I know).

Luckily, there is a great choice for stabilizing mast cells and helping you with your mold toxicity (if your mold is causing MCAS)—ketotifen.

I had to start off really slowly with my ketotifen dosing. In fact, I ended up getting the lowest possible dose they make to begin with—0.1 mg—because it made me so tired at first. However, over time, I have been able to increase to a recommended dose.

According to Woodland Hills Rx Pharmacy (a compounding pharmacy located in California):

Ketotifen is a first-generation H1 antihistamine that is particularly useful for treating mast cell conditions because it also acts as a mast cell stabilizer. It has been used for many years to treat allergic conditions and to prevent asthma attacks. Studies have shown that ketotifen inhibits exocytosis in mast cells and can be effective at low doses of 1 mg.[46]

They go on to say that "the most common prescription we see for adults and older children is 1 mg taken twice daily and for children 0.5 mg twice daily."[47]

If you are interested in taking ketotifen, you will need to have it prescribed by a doctor and ask them about proper dosing for you. Also, make sure it is a fit for your given condition. I have included MCAS in this book because it is often seen in patients with mold toxicity, but you may not have MCAS. A good doctor will ultimately help you determine if you have MCAS or not.

Because ketotifen made me so tired initially (and still can), I only take a higher dose of 1 mg right before bed. In my experience, ketotifen isn't intended to be taken long term, and I've already begun the process of reducing my dose as my body stabilizes and my mold levels decrease.

WHAT IS KETOTIFEN?

"Ketotifen is an antihistamine that reduces the effects of the natural chemical histamine in the body."[48]

"Ketotifen was first developed in Switzerland in 1970 by Sandoz Pharmaceuticals and was initially marketed for the treatment of anaphylaxis ... oral ketotifen is used in Mexico and across Europe for the treatment of various allergic symptoms and disorders, including urticaria, mastocytosis[49], and food allergy."[50]

Ketotifen can be prescribed by a doctor in the United States.

A common next step after getting your bowel movements in proper order is finding a binder that helps clear the mycotoxins you will identify after testing your body for mold.

BINDER OPTIONS

Dr. Neil Paulvin explains that binders help remove toxins from your body by preventing those toxins from passing through the walls of your intestines and into your bloodstream:

> Binders are insoluble particles that can pass through your gut unabsorbed and their role is to attract and bind toxins inside your gut to eventually remove them. Binders can do so since they are positively charged, which attracts a negative charge of mycotoxins.[51]

Binding the toxins means that instead of being reabsorbed into the body, they can pass out of your body in your stool instead.

Of course, that means striking a delicate balance with your digestive system. Not everyone can tolerate heavy binders early on in their detox process. It took me six months to stabilize my body enough by bringing down my histamine levels before I could tolerate binders.

And when you take the binders, you have to make sure that you're having regular bowel movements because that's how the toxins are excreted from your body. Just to make it a bit more complicated, the binders can be, well, binding—they can make you constipated. So you'll need to make sure that you're drinking enough water and taking other steps to promote healthy bowel habits; otherwise all of those bound toxins will still be sitting in your body.

There are many options when it comes to binders. If you know which toxins you have in your body from your testing, it will better help you decide which binder to take, and of course, a good functional medicine doctor will be able to help you choose a binder that best fits your situation.

Binders may include:

- Bentonite clay
- Activated charcoal
- Cholestyramine
- *Chlorella*
- Propolmannan
- *Saccharomyces boulardii*
- Glucomannan
- Fiber
- Probiotics (Make sure you buy a brand that does not contain histamines. I like Seeking Health ProBiota HistaminX.)

You can go to Mosaic Diagnostics website to find a helpful description of many of these binders and a chart that shows which binders go well with eight common mycotoxins. Additionally, you can find information about these binders in the book *Toxic* by Dr. Neil Nathan and on the Mast Cell 360 website.

Beyond these binders, you may also find a Western medicine for mold is helpful. My daughter and son both took nystatin to help clear mold out of their bodies.

Nystatin is an antifungal medication that you can get from a doctor. I haven't been able to take nystatin yet because I am still working on being able to tolerate it, which highlights an important point:

Move slowly!

You must listen to your body as you take this path. You don't want to put further stress and strain on yourself. Baby steps are needed as you heal, and this is an opportunity to honor yourself as you try to detox a small amount at a time (especially in the beginning). Throw away the idea that quicker, bigger, and faster are better. That is not true when it comes to clearing mold toxicity.

Let us return for a moment to Brielle, who has been working diligently to clear the mold from her nine-year-old son.

DETOXING MOLD FROM A CHILD

I asked Brielle (you'll remember her from the case study in Chapter Five) to share what her current protocol is for her son.

In the morning, Brielle gives her son:

- D-Hist Jr. by Springboard (two chewable tablets)
- Fish oil (one teaspoon)

When he has a histamine rash, she gives him:

- Claritin (one capsule)
- Pepcid (one capsule)

At dinner, Brielle gives her son:

- Magnesium (one tablespoon)
- Probiotic *Saccharomyces* (in honey or a spoonful of magnesium)
- N-acetylcysteine (NAC) (in honey or a spoonful of magnesium)

Before bed, Brielle alternates every other day, giving her son:

- Activated charcoal (one capsule—opened into spoonful of honey if unable to swallow)
- Medi-Clay-FX: calcium bentonite clay (one capsule—opened into spoonful of honey if unable to swallow)

If needed, Brielle will also give two pumps of LipoCalm under the tongue when the child seems overstimulated or requires support with winding down before bed.

Since starting this protocol, Brielle hasn't had the opportunity to retest for mold yet and is still in the phase where some of the medicines she is giving cause reactions (this can be normal and goes back to our topic of moving slowly). However, one of the biggest symptoms that Brielle's son had from the mold was nosebleeds and a staph infection that has gone away since moving out of the mold and using nasal spray in conjunction with the protocol above.

It's families like Brielle's that give me great hope for the awareness that is growing around mold and the positive impact detoxing from mold is having on people's lives.

In a recent conversation with another friend who was greatly impacted by mold, he texted me "Thank youuuuuuuuuuuuuuuuuuuuu!" as we talked about the dramatic shift he has had in his well-being since discovering mold six months ago.

Witnessing so many success stories is a breath of fresh air in the otherwise overwhelming problems that mold brings.

I've shared Brielle's protocol with you as a guide and to give you some sense of the kinds of binders, supplements, and medications that are available to support your healing. That said, please remember that this process is not one size fits all. I am not a doctor, and I don't know exactly what your body needs. In sharing these plans, I'm not suggesting that you should follow any of them. Instead, use them as a guide so that when you talk with your doctor, you're well informed on the topic of mold and familiar with some of the strategies being used to help treat mold toxicity.

CALMING TOOLS FOR SUPPORT

By now, your toolbox is growing rapidly, and you have some practices that should be calming your limbic system.

The limbic system includes the hippocampus, the amygdala, and the hypothalamus, and it's the part of your brain that's responsible for processing memories and emotions. It also plays an important part in the fight-or-flight response, and when we keep replaying stressful situations in our mind, it can trigger the same response, even if we're not actively staring down a dangerous situation.

Let's do a few activities that will support your limbic system in preparation for what's to come.

TOOL #1: CREATE POSITIVE EMOTIONS

Sometimes when we are ill, everything can seem heavy. I often think of Eeyore from *Winnie the Pooh*. He always has a sad way about him and struggles to find the good in any situation. When I am feeling like Eeyore, I ask myself what good I can find in the moment, or I try to remember something good that happened recently!

What we focus on grows, and when we focus on the good, we will see more good around us!

In the space below, write three things in your life that you recognize are good—they can be things you're thankful for in the present, heartening recent events, or even things that you can imagine will be good in the near future.

_____

_____

_____

TOOL #2: GRATITUDE BOX

Ready for some arts and crafts? I promise, this isn't as cheesy as it sounds, and the practice can be wonderfully beneficial.

Grab an old shoebox or a cardboard box and some markers and scissors. You can decorate the box however you like, and when you are done, cut a slit in the top of the box for your gratitude writing to go into.

Each day, either start or end your day by writing what you are grateful for on a small slip of paper.

It is easy to skimp on this exercise by always repeating the same things or keeping your gratitude superficial. Don't do that!

This gratitude box is in the self-compassion section for a reason. I would like your gratitude to be focused on what you appreciate about yourself.

Yes, what are you grateful for when it comes to *you*?

This exercise can be a stretch when our bodies don't necessarily feel or look how we want them to. This can be hard when we are feeling depressed, anxious, or like the world is crumbling around us.

So you will be working with self-compassion twofold here. On the one hand, be compassionate as you honor the difficulty of this tool. Know that it isn't easy, and you don't have to do it "exactly right."

Instead, focus on the small things that you can appreciate about yourself. If a dear friend were writing about you, what would they say?

Keep writing what you are grateful for over the next week and then open the box at the end of the week and read each slip of paper to yourself.

How do you feel after reading your gratitude writing?

Can you do this practice for another week?

Can it become ongoing for you?

TOOL #3: WALKING MEDITATION

Whether you carve out time to walk in nature or use walking meditation down the busy streets of New York City, you can bring the same practice and awareness to any space that you are in.

Walking meditation is an invitation to bring attention to your breath and your surroundings. You may try walking slowly with an inhaled breath as you count three steps, followed by an exhaled breath as you take another three steps:

In breath with a Step 1, Step 2, Step 3.

Out breath with a Step 1, Step 2, Step 3.

Other times, you may find yourself walking faster. Even with a quicker pace, you can bring awareness to your breath, your feet as they touch the earth with each step, and the space around you.

Or you may have the opposite experience with walking and find it very difficult. When I was at my sickest, my doctor suggested that I go for a walk every day. I felt that was an impossible task, so I ignored the suggestion completely. Today, I know that I could have practiced walking meditation across my living room or simply walked for two minutes down the street. I have learned that doing a little is better than doing nothing.

Thích Nhất Hạnh (the beloved Zen master and spiritual leader) said, "When you walk, arrive with every step. That is walking meditation. There's nothing else to it."

He also said, "Every path, every street in the world is your walking meditation path."

See if you can begin to incorporate walking meditation into your daily life!

CHAPTER 9

CREATING A HOME FOR HEALING

Imagine that you are on the beach of Costa Rica. Behind you are miles of undeveloped land full of long grasses and trees made for howler monkeys. In front of you is endless sky with a sparkling blue ocean that continues to create perfect wave after perfect wave. You sit and watch the waves roll in—1, 2, 3.

Again, you sit and watch—1, 2, 3.

Imagine that you are strong and healthy, and after you have sat in silent observation for a while, you grab your surfboard and paddle out just beyond the breaking point of the waves.

You inhale the salt of the sea. You exhale as you let your board catch the incoming wave and glide with the flow of the water.

As you paddle back to shore, the sunlight hits your back and you are reminded of all that is good, all that is safe, all that is joyful about this world and your life.

Take a moment to close your eyes to really feel this scene in your body. Let the feelings of safety wash over you.

For many of you who have been sick with mold toxicity for so long, you haven't experienced calming, carefree, or happy moments in a long time.

You most likely feel that you have been getting pounded by the waves rather than surfing with the grace of the waves.

In the words of Jon Kabat-Zinn, "You can't stop the waves, but you can learn to surf."

That, dear reader, is now our work. We must learn how to "surf" in this illness.

LEARNING TO SURF

You may be wondering: *What does it mean to surf in the context of mold toxicity?*

Surfing implies that you are not fighting against your illness. Through testing your home and body or by identifying with the symptoms you have learned about, you accept that you have been affected by mold. You know the source of your illness, and you trust that your body and mind will heal in time.

In order for that healing to take place, though, your nervous system cannot continue to be on high alert all the time. You cannot stand in the break of the waves and let yourself be pounded by the ocean any longer.

No, instead you begin to make peace with your condition and bring in tools that will calm your nervous system so that you are able to take the right binders and medicines required for detoxification.

After years of illness and not knowing how the body will react to a new vitamin or different medicine, it can become very difficult to take new remedies because the body has perceived so much as a threat.

In this chapter, we must take the threat away and begin the process of creating a home for healing.

A HEALING PLACE

My hope is that by now you are in the process of discovering whether you do have mold in your home. I know that testing for mold can take some time while you determine the correct test for you and wait for the results. Hopefully, you have received your results or you will soon.

If your home tests positive for mold, you will find a trained professional to help you remediate the mold while you stay elsewhere, or you will move out of your home if you are renting.

As discussed previously, but something that bears repeating, it is safest for you to not be in a moldy home, and it can be dangerous for you to be in the home during remediation because the exposed mold will release spores into the air as it is being removed.

In my situation, our family rented a smaller house near our children's school during the remediation process.

Thinking about the costs of clearing your home can be intimidating. While we had to renovate our entire house, that's not always going to be the case in every situation. I've had friends who've lived in moldy apartments who were able to renovate their homes for much less.

Because of the costs, you may be tempted to take on the work yourself, but I'd strongly caution you against going that route if you can help it, which is why I'm not covering that topic in this book. As you start to rip out drywall or flooring, you'll be releasing a huge number of additional spores into the air. If you're not properly protected, you'll be putting yourself at risk of breathing those spores in and compounding your exposure; if you haven't properly sealed off the part of the house you're working in, you can wind up spreading those spores throughout your home.

The experts who came into our home to do mold remediation all wore hazmat suits every step of the way—that's how toxic mold truly is.

Even if you can easily afford to hire professionals to take care of your mold problem, though, the process can be draining. In our case, we had to move into temporary housing for more than a year to get all of the work done. It was hard on all of us, and I really had to lean on the self-care tools I had available to get through that time.

Whether you are living in a rental home, staying with friends, or sleeping on your family's couch, you will want to bring the following tools into your space.

It is possible to do this work wherever you are, and the sooner you get started the better, because you deserve to feel healthy!

CALMING TOOLS FOR SUPPORT

Because this is such a proactive phase of your healing, I want to really focus on tools that will help support you during this time. I know that it can feel overwhelming, uncertain, and disruptive. It's hard to recognize that the home you saw as a sanctuary for your family was also making you sick. So these tools are designed to help you develop a renewed sense of trust and peace in your physical space as you begin to heal.

TOOL #1: YOUR HEALING SPACE

When I was a child, my parents owned many acres of land under the big Douglas firs and madrone trees of Mendocino County. We lived there for a short time when I was a baby and then moved closer to the Golden Gate Bridge in California, which connects San Francisco to Marin County.

Although I no longer lived in the woods full time, my mother and I would go up constantly to be on the property and keep it in working order.

We would take the long drive out of the suburbs in my mom's Volvo station wagon, passing the landmarks that I would come to know so well and finally making our way up the dirt road to the cabin my mom had helped build.

Upon arrival, my mom would turn on the cabin's generator, and I would head out of the house to play. With no television or neighbors (and being an only child of divorced parents), it was usually just me outside.

In the garden, I would play with an old iron stove that had been left, pretending I was a chef cooking for my many customers. I would bound down to the river and look for fairies dancing in the chilly water. Sometimes, I would dip my toes in and squeal with delight as the crispness of the water always felt like a surprise (no matter how many times I had done it before).

When I was done in the forest, I would head into our wooden cabin and lie on the futon mattress, looking up at the knots in the wood of the ceiling. I would hunt for shapes in that wood, shapes that reminded me of animals, people, vehicles, or anything else that my mind could conjure up.

That home was a very special space for me. It was a place where I felt the magic of the world, where I felt happy, and it formed many of the core beliefs I have today about the importance of nature, the universe, and magic that I can't always see but can often feel.

Eventually, my mother sold her share of "the land," as we called it, and later my father sold his land as well.

I missed going up to the property, but I never forgot the importance of having that space.

In each of my subsequent homes, I always made sure that I had an area dedicated to the beauty that I had felt when I was a child. It looked different in each one, but its core was the same—reminders of nature, things that I valued, special items, or pictures that represent my beliefs, goals I had, and yes, the magic of the goodness that I have always believed in.

For many this is very similar to an altar, a mediation nook, a positive visualization space, or a shrine.

As my health ebbed and flowed over time, I realized just how important it was for me to create something that spoke to my heart. A space where I could journal, cry, give thanks, and connect with myself when the going got tough.

For this tool, I would like to invite you to create a new space or reexamine an existing space in your home (or wherever you are staying) that will be your spot of safety.

Will you decorate it with crystals and meditation cards?

Will you put up a visualization board with pictures of health and wellness attached?

Will you put in a record player with your favorite songs and a cozy chair to settle into?

Or will you collect several postcards that make you feel happy when you look at them and take them out wherever you are to serve as a reminder of your own personal well-being?

Whatever your space looks like, the important thing is that it makes you feel better. It reminds you of the good that is in your life (even when you aren't or haven't been feeling so well).

Healing is possible, but we must first prepare our bodies to heal by teaching ourselves that it is safe to do so.

I hope you enjoy your new positive space!

TOOL #2: WRITE YOURSELF A LETTER

This tool goes together with Tool #1. Once you have created your healing space, I invite you to write yourself a letter.

Ideally, your letter will be sincere, honest, vulnerable, and optimistic. It is a letter that you will keep and reread as many times as you need to while you sit in your place of healing.

This letter isn't about ignoring what has been true for you up until this moment. We aren't just putting a happy-face sticker over a wound and pretending that nothing ever happened.

No, this letter is as real as it gets. And no one ever has to read this letter but you!

Being sick sucks! Having mold in your home is a real pain (to say the least)! Not knowing *why* you have been suffering for so long feels really unfair sometimes.

Acknowledge all that! Own your pain and write it down. Then, give yourself the reassurance you so desperately need right now.

Here is an example of my letter:

Dear Laura Linn,

First, let me express my sincere apology for all the pain and suffering you went through for decades. It wasn't easy to be sick at such a young age. You often felt alone with your poor health, and you didn't know who to talk to about what was going on for you. In fact, when you did talk about your pain, you often felt judged and disregarded or simply misunderstood.

I'm sorry that you had so many years filled with doctors' appointments, your hair falling out in clumps, painful stomachaches, heightened stress, and uncertainty about your health. I know none of that was easy.

And, yet here you are—a strong and brave woman who hasn't given up. You kept digging for answers when others said there weren't any. You read countless books about health and nutrition, sought out expert advice, looked at everything from emotional support to physical support. And all of those pieces made up different parts of the puzzle. A puzzle, which I must admit, is beginning to look better each and every day.

I am so proud of all your hard work. I am so proud of the woman you have become today. Life gave you lemons, and you turned those lemons into a lemonade stand where you could give it away (and drink some yourself!).

Your pain helps others. Your pain makes you more empathetic. Your pain has taught you to take less for granted and find more compassion in this human experience that we are having.

In some ways, it almost seems fitting to thank your pain. To thank it for all it has taught you over the years. In thanking your pain, you can also give yourself (and it) permission to let it go.

You no longer need to hold on to that pain because you are getting to the root of what was causing your poor health. You are coming into a new and healthier stage of life. There is so much to be grateful for.

First and foremost, let's start that gratitude for the healthy life that is coming your way!

In the words of Maya Angelou, "We delight in the beauty of the butterfly, but rarely admit the changes it has gone through to achieve that beauty."

I see your changes and the beauty that is coming to fruition as you go through your metamorphosis.

Congratulations and tons of love,

Laura Linn

Notice how this letter is written with me talking to myself in third person. You can experiment to find what feels best for you. I find that writing in the third person makes the letter feel like it is coming from a best friend, and I find a lot of comfort in that.

Keep your letter in your special place that you created in Tool #1, and read it as many times as you need.

TOOL #3: LETTING GO OF OLD STORIES

There is an old Buddhist parable that I would like to paraphrase and expand upon for you.

In the parable, a junior monk and a senior monk are walking together when they come upon a young woman who is standing at the bank of a river. With the strong current, the woman is unable to cross the river alone.

Although the monks have taken vows not to touch a woman, the older monk sees her need for help and carries her safely across the river and sets her down on the other side.

Then, the two monks continue on their path again.

As the two monks walk, the junior monk is speechless. He is deeply bothered that the senior monk broke his vow.

An hour passes and then another hour. Finally, after several hours the junior monk feels he must say something.

The junior monk turns to the senior monk and says, "How could you break your vow and carry that woman across the river?"

The senior monk looks the junior monk in the eye and responds, "I put her down hours ago by the side of the river. Why are you still carrying her?"

For many affected by mold toxicity, we are unconsciously carrying our story of sickness with us everywhere we go.

It is not that we shouldn't honor the reality of our situation (we absolutely should, and that is why Tool #2 is so important), but we don't want to cling to our story so tightly that we miss the beauty of the walk like the junior monk did.

When you notice your mind starting to wander off into worries about your health and ruminating on past stories of shame and guilt that surround your illness, I invite you to actively pause with your breath.

Take a deep inhale through your nose for a count of 1, 2, 3.

Hold your breath briefly at the top of your inhale and then let out an audible exhale for a count of 1, 2, 3.

Again, inhale through your nose for a count of 1, 2, 3.

Hold your breath briefly at the top of your inhale and then let out an audible exhale for a count of 1, 2, 3.

Now, bring in this mantra (or make one up that is fitting for you):

I am healing; I am safe; I am letting go of past worries and judgments.

Again, tell yourself:

I am healing; I am safe; I am letting go of past worries and judgments.

Can you leave your story by the side of the river like the senior monk did? It is possible when you actively work on it through breath, mantras, and meditation.

TOOL #4: DAILY JOURNALING

Journaling can:

- Reduce anxiety and depression
- Improve memory and comprehension
- Help boost our immune system
- Help with self-confidence
- Help with recovery from trauma
- Lead to better sleep

With all the benefits of journaling, it is no wonder that I sneak in journal prompts throughout the book as much as possible. The benefits are pretty amazing, right?!

So, now that you know just how powerful journaling can be, I invite you to grab a notebook or staple some pieces of paper together for the following daily journal prompts.

Let's give this a try for thirty days and see how we feel after we have done it for a month! Deal?!

Journal Prompt #1: What are three things in my life that I can look forward to today?

Journal Prompt #2: What can I do today to decrease my stress?

Journal Prompt #3: Today, I am grateful for _____.

Journal Prompt #4: My mantra for today is _____.

Journal Prompt #5: Imagine what it will feel like to be in good health. Will you have extra energy? Will you be able to go back out into your garden? Will you have more time to connect with your children? Hold that vision of being healthy in your mind, and then write down the positive feelings you experience in the space below.

_____

_____

_____

_____

All these tools/practices can be done in the safe space that you are creating within your home. Notice how things begin to shift as you take these actions toward creating more health and happiness in your life.

LAST WORDS

I t is the summer of 2022. I don't know who is currently popular on the radio, but Michael Franti & Spearhead are singing on my playlist as I write this.

The song "Good Day for a Good Day" pulses through my phone as I write these last words, and it seems very fitting.

I no longer have stomach pain. I can do yoga for the first time without getting adrenal fatigue. I am no longer exhausted. I have clarity and peace of mind, and I feel strong and healthy.

Yes, you read that right, friends.

I feel strong and healthy!

Gratitude would be an understatement. In all honesty, I may have run out of words to describe how I feel. So, in this case I will borrow words to describe how I feel from great Mother Teresa:

"Peace begins with a smile."

Today, my internal peace begins with a smile, and I share that smile with you.

I know that healing is possible. It was possible for my daughter, it is possible for me, and it is possible for YOU.

As we part from this book, please accept the joy I pass along through these words, the peace that will continue to manifest in your life, and please hold on to the truth that your light is meant to shine bright!

I smile for your healing and for the joy it will bring to you and those around you.

RESOURCES

In addition to this list, you can find more information, additional resources, and direct links to these books and websites on my site: https://thetoxicmoldsolution.com.

UNDERSTANDING THE SCIENCE OF MOLD AND MOLD TOXICITY

Lilley, James. *Toxic Mold Book: How to Remove Mold and Mycotoxins from Your Home.* Self-pub, 2020.

Nathan, Neil. *Mold and Mycotoxins: Current Evaluation and Treatment.* BookBaby, 2022.

Nathan, Neil. *Toxic: Heal Your Body from Mold Toxicity, Lyme Disease, Multiple Chemical Sensitivities, and Chronic Environmental Illness.* Las Vegas, NV: Victory Belt Publishing Inc., 2018.

Shippy, Ann. *Mold Toxicity Workbook: Assess Your Environment & Create a Recovery Plan.* Austin, TX: Treaty Oak Publishers, 2016.

TESTING YOUR HOME FOR MOLD

ImmunoLytics: https://immunolytics.com
ImmunoLytics offers DIY gravity-plate and swab testing. They also give you the option to self-analyze your gravity plates or send them to a lab and receive a detailed report and a consultation with an expert. Their swab tests include lab fees and a report since there's no way to self-analyze those samples.

Lis Biotech: lisbiotech.com
Lis offers DIY ERMI test options, but if that's cost-prohibitive for you, they also offer DIY HERTSMI testing kits.

TESTING YOUR BODY FOR MOLD

International Society for Environmentally Acquired Illnesses (ISEAI) database: https://iseai.org/find-a-professional
Having an expert partner in this process is so beneficial. The ISEAI database is the best resource for locating a functional medicine doctor who can help guide you and order the appropriate tests for your unique situation.

RealTime Laboratories, Inc.: https://realtimelab.com
RealTime Lab offers mycotoxin urine testing. They also host a blog with helpful information about mold and mold exposure, which you can find on their website.

Mosaic Diagnostics: www.mosaicdx.com
Mosaic Diagnostics offers both mycotoxin testing and organic acids tests. Their website also has an extensive resources page with articles, information, and even webinars.

HEALING YOUR BODY

Mast Cell 360: mastcell360.com
Beth O'Hara, a functional naturopath, has put together an incredible resource with information about MCAS, including a blog and several courses.

Sinusitis Wellness Blog: www.sinusitiswellness.com/blog
Don't let the name fool you—this blog covers a wide range of topics on mold and mold toxicity, including supplements, detoxification, fatigue, and digestive health.

Well-Hacking Tips from Dr. Paulvin: doctorpaulvin.com/blog
Neil Paulvin's blog has several great articles about mold toxicity
and treatments.

ENDNOTES

1 Please know that when I use the term "anxiety" in this book with reference to my daughter's experiences, I am doing so in a nonclinical sense to describe feelings of apprehension, distressing concern, unease, nervousness, or dread, not as a diagnosis, medical or otherwise.

2 Neil Nathan, *Mold and Mycotoxins: Current Evaluation and Treatment* (BookBaby, 2022).

3 Neil Nathan, *Toxic: Heal Your Body from Mold Toxicity, Lyme Disease, Multiple Chemical Sensitivities, and Chronic Environmental Illness* (Las Vegas, NV: Victory Belt Publishing Inc, 2018).

4 "What Is Mold Illness? Better Yet, Do People Get Sick After Being Exposed to Water-Damaged Buildings?" Surviving Mold, accessed October 27, 2022, www.survivingmold.com/resources-for-patients /diagnosis.

5 James Lilley, *Toxic Mold Book: How to Remove Mold and Mycotoxins from Your Home* (self-pub, 2020).

6 As quoted in "How to Identify Hidden Mold Toxicity (and What to Do About It)," *Goop*, accessed October 7, 2022, https://goop.com /wellness/health/how-to-identify-hidden-mold-toxicity-and-what-to -do-about-it.

7 Nathan, *Toxic*, 41.

8 Nathan, *Mold and Mycotoxins*.

9 Centers for Disease Control and Prevention, "Basic Facts About Mold and Dampness," Mold: General Information, last updated September 13, 2022, www.cdc.gov/mold/faqs.htm.

10 J. D. Spengler et al., "Respiratory Symptoms and Housing Characteristics," *Indoor Air: International Journal of Indoor Environment and Health*

4, no. 2 (1994), 72–82, https://dx.doi.org/10.1111/j.1600-0668.1994 .t01-2-00002.x.

11 "The Top 10 Worst States for Mold," HomeAdvisor, last updated March 2, 2017, www.homeadvisor.com/r/10-worst-states-for-mold.

12 Quest Diagnostics, "Allergies Across America: The Largest Study of Allergy Testing in the United States," published 2011, https://healthland.time.com/wp-content/uploads/sites/4/2011/05 /2011_qd_allergyreport.pdf.

13 Allergy Asthma Food Allergy Centers, "Seasonal and Year-Round Allergies (Allergic Rhinoconjunctivitis)," published January 27, 2021, https://aafacenters.com/seasonal-allergic-rhinitis.

14 Neil Paulvin, "Mold 101: Mold Toxicity," *Well-Hacking Tips from Dr. Paulvin*, last updated September 28, 2020, https://doctorpaulvin.com /blog/mold-101-mold-toxicity.

15 Nathan, *Mold and Mycotoxins*.

16 Federal Emergency Management Agency, "Dealing with Mold & Mildew in Your Flood Damaged Home," accessed December 7, 2022, https://www.fema.gov/pdf/rebuild/recover/fema_mold_brochure _english.pdf.

17 Jon Kabat-Zinn, "Mindfulness-based Interventions in Context: Past, Present, and Future," *Clinical Psychology: Science and Practice*, 10 (2), 144–156.

18 Nathan, *Toxic*, 49–52.

19 Saul McLeod, "Stress, Illness, and the Immune System," *Simply Psychology*, last updated 2010, https://www.simplypsychology.org/stress -immune.html.

20 "Chronic Stress Puts Your Health at Risk," Mayo Clinic, published July 8, 2021, https://www.mayoclinic.org/healthy-lifestyle/stress -management/in-depth/stress/art-20046037.

21 Stephanie Hairston, "How Grief Shows Up in Your Body," WebMD, published July 11, 2019, https://www.webmd.com/special-reports/grief -stages/20190711/how-grief-affects-your-body-and-mind.

22 Jon Kabat-Zinn, "Mindfulness Body Scan by Jon Kabat-Zinn," Mindfulness Training, accessed October 7, 2022, https://mbsrtraining .com/mindfulness-body-scan-by-jon-kabat-zinn.

23 Jane Sandwood, "Could Mold Be Affecting Your Mental Health?," *Mental Health Connecticut Blog*, published March 26, 2020, https://www .mhconn.org/mind-body-health/could-mold-be-affecting-your-mental -health.

24 "Could Toxic Mold Cause Mental Health Troubles?," *Polygon Blog*, accessed October 7, 2022, https://www.polygongroup.com/en-US /blog/could-toxic-mold-cause-mental-health-troubles.

25 There have been numerous studies linking mold exposure and neurological and psychological symptoms. They can be dense and challenging to sort through, but they're also full of useful information if you're looking for hard data from the medical and scientific community. See, for example, Cheryl F. Harding et al., "Mold Inhalation Causes Innate Immune Activation, Neural, Cognitive, and Emotional Dysfunction," *Brain, Behavior, and Immunity* (July 2020), 218-228; Aarane M. Ratnaseelan et al., "Effects of Mycotoxins on Neuropsychiatric Symptoms and Immune Processes," *Clinical Therapeutics* 40, no. 6 (June 2018), 903-917, https://doi.org/10.1016/j.clinthera.2018.05.004; L. D. Empting, "Neurologic and Neuropsychiatric Syndrome Features of Mold and Mycotoxin Exposure, *Toxicological and Industrial Health* 25, no. 9 (October/November 2009), https://doi.org/10.1177 /0748233709348393; German Torres et al., "Indoor Air Pollution and Decision-making Behavior: An Interdisciplinary Review," *Cureus* 14, no. 6 (June 2022), https://doi.org/10.7759/cureus.26247; Carol Potera, "Mental Health: Molding a Link to Depression," *Environmental Health Perspectives* 115, no. 11 (November 2007), https://doi.org/10 .1289%2Fehp.115-a536a; B. R. Crago et al., "Psychological, Neuropsychological, and Electrocortical Effects of Mixed Mold Exposure," *Archives of Environmental & Occupational Health* 58, no. 8 (August 2003), 452-63; Cheryl F. Harding et al., "Mold Inhalation Causes Innate Immune Activation, Neural, Cognitive, and Emotional Dysfunction," *Brain, Behavior, and Immunity* 87 (July 2020), https://doi .org/10.1016%2Fj.bbi.2019.11.006; W. A. Gordon et al., "Cognitive Impairment Associated with Toxigenic Fungal Exposure: A Replication and Extension of Previous Findings," *Applied Neuropsychology* 11, no. 2 (2004), 64–75.

26 Here's an explanation of "pure O" OCD that I find really helpful: "Just like other types of OCD, people with pure O have obsessions and compulsions. But the difference is, most (or all) of the compulsions are mental. This means the compulsions are invisible to other people." Sarah Gupta, "What Is "Pure O" OCD?," GoodRx Health, published September 17, 2021, https://www.goodrx.com/conditions/obsessive -compulsive-disorder/what-is-pure-o-ocd.

27 "ERMI Testing Lab Services," Eurofins, accessed October 7, 2022, https://www.emlab.com/services/ermi-testing.

28 "Using the DIY Mold Test Kit: Step 4, Prep for Analysis, Optional: Self-Analysis," ImmunoLytics, accessed April 17, 2023, https://immunolytics.com/mold-testing/mold-testing-guide.

29 Bill Weber, email message to the author, June 5, 2022.

30 J. R. Girman et al., "Prevalence of Potential Sources of Indoor Air Pollution in U.S. Office Buildings," Environmental Protection Agency, Proceedings: Indoor Air 2002, https://www.epa.gov/sites/default /files/2014-08/documents/base_4c301.pdf.

31 General Accounting Office, *School Facilities: Condition of America's Schools*. 1995, General Accounting Office: Washington, DC, https:// www.gao.gov/products/HEHS-95-61.

32 Nathan, *Toxic*, 49.

33 Nathan, *Toxic*, 50.

34 "What Is Functional Medicine?," Functional Medicine SF, accessed October 7, 2022, https://functionalmedicinesf.com.

35 "What Is Naturopathic Medicine?," Association of Accredited Naturopathic Medical Colleges, accessed July 13, 2023, https://aanmc .org/naturopathic-medicine.

36 Nathan, *Toxic*, 61.

37 For more information on the effects of macrocyclic trichothecenes, see "Trichothecene Information," RealTime Laboratories, Inc., accessed April 17, 2023, https://realtimelab.com/trichothecenes.

38 "What Are Organic Acids?", Great Plains Laboratory, LLC, accessed October 7, 2022, https://www.greatplainslaboratory.com /organic-acids-test. Please note that Great Plains Laboratory became Mosaic Diagnostics while I was in the process of writing this book. This information came from their previous website.

39 Susan Tanner and Eric Hosford, "Why Gluten and Yeast Can Trigger an Endless Cycle of Reactivity for Mold Sufferers," *Sinusitis Wellness*, published May 24, 2021, https://www.sinusitiswellness.com /why-gluten-and-yeast-can-trigger-an-endless-cycle-of-reactivity -for-mold-sufferers.

40 "The Brain-Gut Connection," Johns Hopkins Medicine Health, accessed October 7, 2022, https://www.hopkinsmedicine.org/health /wellness-and-prevention/the-brain-gut-connection.

41 Remember, an all-white diet can exacerbate your symptoms if you have mold toxicity.

42 There are many others as well; I recommend the list Beth O'Hara, a functional naturopathy doctor, shares on her blog, *Mast Cell 360*. You can find the URL for her site in the resource guide at the end of the book.

43 Beth O'Hara, "What Is MCAS? Mast Cell Activation Syndrome Basics," *Mast Cell 360*, accessed October 7, 2022, https://mastcell360 .com/what-is-mcas.

44 Beth O'Hara, "Testing Mycotoxins and Mold: One of the Biggest Root Triggers for Mast Cell Activation and Histamine Intolerance," *Mast Cell 360*, accessed October 7, 2022, https://mastcell360.com /mycotoxins-and-mold-one-of-the-biggest-root-triggers-for-mast -cell-activation-histamine-intolerance.

45 Jan Chozen Bays, "Mindful Eating: How to Really Enjoy Your Meal," Psychology Today, published February 5, 2009, https://www .psychologytoday.com/us/blog/mindful-eating/200902/mindful -eating.

46 "Ketotifen for Mast Cell Activation Syndrome (MCAS)," *Park Compounding Pharmacy Blog*, accessed October 7, 2022, https://www .woodlandhillspharmacy.com/ketotifen-mast-cell-activation-syndrome.

47 "Ketotifen for Mast Cell Activation Syndrome (MCAS)."

48 Cerner Multum, "Ketotifen ophthalmic," Drugs.com, published April 29, 2022, https://www.drugs.com/mtm/ketotifen-ophthalmic .html.

49 For reference, urticaria is the itchy red welts we associate with allergic reactions. Mastocytosis is a condition in which the overproduction of mast cells results in those cells collecting in the body's tissues, like the skin and internal organs, which sometimes causes systemic damage.

50 "Ketotifen," Drug Bank Online, last updated October 7, 2022, https://go.drugbank.com/drugs/DB00920.

51 Neil Paulvin, "Mold's Effects on Human Health & Treatments," *Well-Hacking Tips from Dr. Paulvin*, last updated October 29, 2020, https://doctorpaulvin.com/blog/mold-101-mold-toxicity.

ACKNOWLEDGMENTS

First and foremost, I would like to recognize and thank all the people who shared their stories for this book so that we could learn from their experiences and heal. Thank you for taking the time to be vulnerable, honest, and informative. Your voices are so very needed!

I have been blessed to have a loving partner with me on this journey. My husband, Tyler, supported me as our family healed from mold toxicity. Thank you, honey, for being on countless hours of phone calls with inspectors, doctors, and mold specialists. Clearing our homes and our bodies of mold has been no small feat, and I greatly appreciate the time and energy you have spent walking this healing path with me.

I am so thankful for all the healthcare professionals and experts in the field who have supported our family directly! Thank you for educating yourselves so that you could educate and heal us. You are making a huge impact on people's mental and physical well-being.

My family and friends have given endless support as they loved me through this healing process. Having a nurturing community around me is miraculous. Special thanks to my dad and stepmother for caring for and about me when my health was at its worst and for being a source of love and kindness through this journey.

I am especially grateful for my mama, Kathy, for being beside me every step of the way on this medical roller coaster. Mama, you have been with me through countless doctors' appointments, have always helped me research healthy choices, and were available for each and every tummy ache, migraine, and moment that I needed you. I am blessed to have you!

My love, pride, admiration, and gratitude for my two children, Oliver and Grace, is on every page and in every word of this book. You have both been so courageous in walking your paths toward healing. Your journey has already helped so many other children and families, and your experience will continue to support and educate those who are learning to heal and become well again. Thank you for allowing me to share your stories.

And last but not least, thank you, dear reader. Thank you for taking the time to buy this book, to educate yourself, and to take action so that you can heal yourself and your family. Knowledge is power, and you are empowering yourself, and thus, your community. We need people like you to share this information so that others can heal as well. Thank you!

ABOUT THE AUTHOR

Laura Linn Knight is a parenting educator, author, mindfulness and meditation guide, mother of two, and former elementary school teacher who helps families create calmer homes by giving them the tools and resources they need to heal.

Today Laura is drawing on her years of experience in family education to support families in addressing a widespread but misunderstood health concern: mold toxicity. Laura shares her knowledge, research, and personal experiences in recovering from mold exposure through her advocacy work and her blog, *The Toxic Mold Solution Blog*.

Laura is the author of *Break Free from Reactive Parenting*, and her work has been featured on NBC's *Today Show*, Romper, PureWow, Motherly, *Good Day LA*, and across other various media outlets.